ORTHOPAEDIC RESEARCH IN SMALL ANIMALS

NIPA GENX ELECTRONIC RESOURCES & SOLUTIONS P. LTD.
New Delhi-110 034

ORTHOPAEDIC RESEARCH IN SMALL ANIMALS

Swapan Kumar Maiti

Principal Scientist
Division of Surgery
ICAR-Indian Veterinary Research Institute
Izatnagar, Uttar Pradesh, India

NIPA GENX ELECTRONIC RESOURCES & SOLUTIONS P. LTD.
New Delhi-110 034

NIPA GENEX ELECTRONIC RESOURCES & SOLUTIONS P. LTD.

101,103, Vikas Surya Plaza, CU Block
L.S.C.Market, Pitam Pura, New Delhi-110 034
Ph : +91 11 27341616, 27341717, 27341718
E-mail:newindiapublishingagency@gmail.com
www: www.nipabooks.com

For customer assistance, please contact
Phone: + 91-11-27 34 17 17 Fax: + 91-11- 27 34 16 16
E-Mail: feedbacks@nipabooks.com

ISBN: 978-93-90591-55-8

Composed and Designed by NIPA.

Dedicated to my beloved Mother

Maa is my first and greatest teacher
a teacher of compassion and love
If love is sweet as a flower, then
my Mother is that sweet flower of love

Preface

This book covers most of the topics with latest information on bone in general and research on bone healing in particular. The book is divided into 27 chapters covering almost all aspect of orthopaedic research in small and lab animals including basic knowledge about bone; fracture types and fracture healing; bone grafts and bone substitute; internal fracture fixation; external fracture fixation; bone morphogenetic protein; transforming growth factors; role of mesenchymal stem cell in osteoinduction; fracture healing in critical sized bone defect and in very large bone defect, effect of herb and herbal product in fracture healing, role of different hormones, anabolic steroid, NSAID drug, bone wax, ultrasound in fracture healing; minimal invasive fracture repair; management of comminuted femoral and tibial metaphyseal fracture; endocrine role of fracture healing; evaluation of bone graft, ceramic biomaterials in fracture healing and physiotherapy of orthopedic patients. This book includes different research findings on application of herb, herbal product, bone graft, ceramic biomaterials, mesenchymal stem cells, and different osteoinducers in bone regeneration. The main objective of this book is to provide the latest information to meet the requirements of not only undergraduate and post graduates research scholars but also to the teachers, biologists and clinician involved in animal treatment and orthopaedic research. The book contains more than 150 good quality photographs of internal fixation techniques, x-ray, undecalcified ground sections, scanning electron microscopy, 3-D CT scan, histopathogical stained sections used in different orthopaedic interventions.

This book would be of immense use to the students, teachers and research scholars engaged in the field of orthopaedic research.

Author

Acknowledgements

I am grateful to the Director and Joint Directors of Indian Veterinary Research Institute, Izatnagar (UP) for their generous help and support towards this publication. I wish to humbly acknowledge the respected authors and publishers of different books/papers consulted and reproduce some text /photos for the preparation of this task with special mention to JA Buckwalter, MJ Glimcher, RR Cooper, R Recker, J K Roush, Hamish Denny, Mark C Rochat, John T Payne, R M McLaughlin, Nanumaker DM, H Rosen, RL Rudy, CC Hudson, P Antonia, DD Lewis, Kalyani Publishers; Martens A, Chaudhari P, Muller M, Gary C Jones, Richard Marsell, Bromiley MW, Thomas A Einhorn, Richardson DW, Deyoung DJ, Ehner EA, Urist MR, Ortved KF, RKC Bali. If I missed any name, please pardon me.

I wish to acknowledge and convey my sincere regards to my mentor and my beloved teacher Dr Gaj Raj Singh, who taught me the first lesson of bone and orthopaedic research. I also wish to acknowledge to my UG teacher Dr Ranjit Ghosh who taught me the bone anatomy; from his class-note, I have taken some information for this book.

Help and wholehearted support extended by all the faculty members in general, especially Drs Naveen Kumar, Kiranjeet Singh, Aswathi Gopinathan and my most beloved PG students of Surgery Division, IVRI especially Drs Sujit Suryawanshi, B Saravanan, Avijit Dutta, Bhagyashree K Bhadane, Ninu A R, Shiva Kumar M U, Sandeep Kumar, Sandeep Khimta, Dileep Kumar, Ajith P, Tamil Mahan, Sangeetha P, Divya Mohan, Shivaraju S, Kalaiselvan E and Mamta Mishra in particular to bring out the publication are duly acknowledged. I have no hesitation to acknowledge their hard research work which is the backbone of this book; actually they are the Co-authors of this book.

I gratefully acknowledge to Department of Biotechnology (DBT) for providing necessary funds and facility to author for carry out stem cell research under approved research project as Principal Investigator; some research results included in this book.

Mere coining of words is not enough for encompassing the whole hearted support and sacrifice of my beloved wife Rina and sons-Akashneel and Abhinandan, my brothers-Tapan and Anup, sister-Smritikana, sister-in-law-Rita Roy, Gopa Roy and Dipa who gave me all the strength and courage to accomplish this arduous task.

I would appreciate to receive any suggestion and constructive criticism about the publication and my efforts would be to further improve more subject and latest information in the next edition. Hope my efforts will help to the clinicians, researchers and students to alleviates the suffering of animal pain through orthopaedic technology.

Date: 6th August, 2020 **Swapan Kumar Maiti**

Contents

1

Bone-Basic Consideration

Bone is an integral part of musculo-skeletal system of body. In the living animal it is bluish pink, whereas, fresh dead bone is yellowish-white in colour. It is hard and resistant to pressure but also have certain amount of elasticity. Its tensile strength is equal to the cast iron, though bone three times lighter and ten times flexible to cast iron.

Number of bones: The number of bones of the skeleton varies with age and species. Number of bones in cattle-210; Horse-208; Dog-271; Fowl-166 and human-208 (207-212)

Composition of bone: Bone contains different bone cells, organic and inorganic matrix and water. Organic matrix is about one-third and Inorganic matrix is two-third weight by volume. The organic material consists of 90% collagen [type-I (maximum), type V and type XII] and 10% non-collagenous glycoprotein and bone specific proteoglycans (osteocalcin, osteonectin, sialoprotein, bone phosphoproteins and small proteoglycans) and bone growth factors [TGF-beta family; Insulin-like growth factor-1, 2; bone morphogenetic protein (BMP); platelet derived growth factor (PDGF); Interleukin -1 (IL-1); Interleukin-6 (IL-6); colony stimulating factor]. Organic matrix of bone resembles the matrix of dense fibrous tissues such as tendons, ligament and joint capsule. It is not certain whether all of these proteins are synthesized by bone cells or whether they are synthesized by cells outside of the bone and then incorporated into the bone matrix. However, their presence within bone and their potential to affect the activity of bone cells suggests they have important role in controlling bone cell functions. Inorganic matrix of bone or mineral component of bone serves as an ion reservoir and gives bone of its stiffness and strength. It reserves 99% of body calcium, 85% body phosphorus and 40-60% body sodium and magnesium. Inorganic matrix of bone is composed of 85% calcium phosphate, 12% calcium carbonate, 3% of magnesium phosphate, calcium fluoride and magnesium carbonate. It also contains traces of chlorine, sodium, potassium. These minerals remain as hydroxyapatite crystal form within bone. However, bone mineral crystal is not pure hydroxyapatite, instead they contain both carbonate and acid phosphate groups (HPO_4) and bone crystal does not contain OH groups. Bone matrix has such great durability and stability that it can remain essentially unchanged and retain much of its strength for centuries after death. Removal of organic matters renders the bone fragile, brittle and shattered, whereas, decalcification renders bone soft and pliable and resistant to fracture.

Types of Bone: bones are classified on the basis of their general shape and function into five groups. These are (i) long bone, (ii) flat bone, (iii) short bone, (iv) irregular bone and (v) pneumatic bone. Long bone (Humerus, tibia, femur, radius ulna etc) have an elongated cylindrical part, called as shaft or diaphysis and two expanded extremities called as epiphysis. The shaft encloses the medullary cavity which in life time is filled up by bone marrow. The wall consists of dense, compact tissues of considerable thickness in the middle part of the shaft of the bone, but becoming thinner towards the extremities. Inside this compact bone, there is thin layer of spongy bone lining the medullary cavity. Each epiphysis consists of spongy bone covered by a thin layer of compact bone. Long bone is found in the limbs and act as pillars of support and as levers. Flat bones- are plate-like bones affording large surface area for the attachment of muscles and protection of adjacent structures. The flat bones (e.g. Scapula) are made up of varying thickness of spongy bones interposed between the two layers of compact bone. Short bones (carpals, tarsal) are made up of a thin layer of compact tissue outside and spongy tissues inside. Sesamoids bones are short bones which do not bear the weight of the body. Largest sesamoids bone in the body is patella. Irregular bone (vertebrae) is irregular in shape and is usually placed on the median line of the body. They resemble short bone in structure. Pneumatic bone contains air space within the compact substance instead of spongy bone and marrow. These cavities are called as sinuses and are lined by mucous membrane. They communicate indirectly with external air. In the fowl, many of the bones are pneumatic (vertebrae, sternum, humerus, femur) and these receive air through the medium of "air sacs".

Covering of Bone: Bone is cover by a membrane termed as periosteum, which covers the outer surface of bone except in the regions immediately around or within synovial joint, at the sites of tendon, ligaments and intraosseous membrane insertions. The periosteum consists of two layers: an outer layer that is dense and fibrous and inner layer that is looser and more vascular and cellular. The inner layer contains cell that are capable of becoming osteoblasts and referred to as Cambium or osteogenic layer. The cells of this layer can also form hyaline cartilage under appropriate circumstances, and they help to form extra-osseous callus during fracture healing referred as external or periosteal callus. The periosteum changes with age. The thick cellular Cambium layer in young animals readily forms new bone. With increasing age, the periosteum becomes thinner and its osteogenic capacity decreases. Despite these changes, periosteal cells continue to form new bone throughout life. Endosteum is a thin membrane lining the medullary cavity and larger Haversian canals.

Bone marrow: There are two kinds of bone marrow- red and yellow. Red marrow occupied the interstices of spongy bone everywhere and medullary cavity of long bones at birth. Red marrow is an important blood forming substances and contains precursors of erythrocytes and granular leucocytes of the blood in addition to Giant cells which give rise to platelets and few fat cells. After birth the red marrow is gradually replaced by yellow marrow which consists of ordinary adipose tissues. Yellow marrow fills the spaces of the spongy bone, short bones and medullary cavity of long bones in the adult, but red marrow persists in the vertebrae, sternum, ribs and skull-bones.

Development of bone: In the embryo the future elements of the skeleton are derived from the mesenchyme which is derived from the Mesoderm. The flat bones of the skull are developed by ossification directly from the membranous layer. This process of ossification is formed by the condensation of mesenchymal tissues; the process is termed as the intramembranous method of ossification. In all the other bones (appendicular skeleton, vertebral column, ribs etc) cartilage is laid down in these membranous layers and destroyed before ossification takes place, the process being termed as Intracartilaginous or Endochondral method of ossification. The process of ossification does not take place simultaneously all over the cartilaginous or membranous representatives of a future bone. It begins at one or more points, known as center of ossification, from which it extends until the entire soft tissue is replaced by bone. In a long bone, there is a primary center of ossification for the diaphysis and secondary centers appear for the epiphyses. Additional centers may appear in some bones for the various processes. The number of centers is different for the various bones in the body but the number for each bone is constant in a particular species.

Blood supply to the bone: Mature bones have an elaborate vascular system that supplies the cells of the marrow, bone tissue and periosteum. Disruptions of the blood supply to the bone due to disease, injury or operative procedure can cause necrosis and impair healing. Despite the anatomical differences, all long bones have the same general pattern of blood vessels, consisting of two circulatory systems; a) Periosteal-diaphyseal-metaphyseal system and b) epiphyseal-physeal system. These two systems form anastomoses of variable extent on and within the periosteum and across the physes. The diaphysis and metaphysis of long bones have three sources of blood supply: a) Nutrient arteries, b) Epiphyseal and Metaphyseal arteries and c) Periosteal arteries. Nutrient arteries pass through the nutrient foramen of the diaphysis and branch proximally and distally within the medullary cavity. Proximal and distal branches of the nutrient arteries join multiple fine branches (anastomosis) from the periosteal and metaphyseal arteries to form medullary arterial system. Under normal circumstances, this medullary vascular system supplies most of the periosteum-covered bone, therefore the primary direction of blood flow through the cortex is centrifugal. Nutrient/Medullary arteries supplies 2/3 of cortex, whereas outer 1/3 cortex is supplied by periosteal arteries. The periosteum also has an elaborate vascular system. A vascular plexus lies over the fibrous layer, cambium layer and outer 1/3 rd cortex and anastomoses with nutrient artery. It is also anastomoses with the vessels in skeletal muscle. With increase age, the number of periosteal vessels diminishes, their contribution to osseous blood supply may decline, but the periosteal vascular network remains as important part of osseous circulation throughout life. The periosteal vessels may also be an important source of blood supply to skeletal muscle. The blood supply to the both layer of periosteum also depends on vascular connection with skeletal muscle. Periosteum can survive and form bone after it has stripped from the bone, as long as the vascular connections between the muscle and the periosteum remain intact. This dual circulatory system explains why diaphyseal and metaphyseal bone can remain viable and fractures can heal after either medullary reaming or periosteal stripping. It also explains why segmental fractures that disrupt the medullary vascular supply in a limb that also has extensive soft-tissue injury or periosteal stripping may result in delayed union or non-union that may damage both the medullary and the periosteal blood supply. Epiphyseal-physeal blood supply: The blood supply to most epiphyses is more precarious than the supply to the periosteum, diaphysis or metaphysis. In immature bone, epiphyseal artery supply the epiphysis and growth plate and it anastomoses with periosteal and metaphyseal artery at the periosteal side. It cannot cross the growth plate. The major difference between the blood supplies of a mature and an immature long bone stems from the inability of blood vessels to cross the physis (cartilaginous growth plate), which separates the epiphyses from metaphysis. The epiphyses and the metaphysis, therefore, have separate blood supply. With the maturity of physes, it closed, not exist and the entire expanded end of the long bone become the metaphysis- which is supplied by metaphyseal and periosteal artery.

Structure of bone: (a) Compact or Dense bone-white and hard. Histologically, the compact substance is arranged in the form of concentric lamellate pattern known as Haversian system. (b) Cancellous or Spongy bone or Trabecular bone- interlacting partition forming a mesh work with spaces containing marrow. It is presence in epiphysis of long bone, flat bone, and short bone. Both cortical and cancellous bone has same matrix composition and structure, only differences in distribution and arrangement of matrix. The mass of the cortical bone matrix per unit of volume is much grater than cancellous bone. Cortical bone has greater density, less porosity (approx 10%), whereas, cancellous bone has lesser density, more porosity (50-90%). Comprehensive strength of cortical bone is approx 10 times greater than those of similar volume of cancellous bone. Cortical bone forms approx 80% of the mature skeleton. The structure of both cancellous and cortical bone changes in response to applied loads, immobilization and hormonal factors. Cancellous bone has approximately 20 times more surface area per unit of volume than does cortical bone. Cancellous bone contain less bone cells lies between lamellae or in the surface of the trabeculae, whereas, cortical bone contain more bone cells and directly influenced by adjacent bone marrow cells. Cancellous bone usually has a higher rate of metabolic activity and remodelling and appears to respond more rapidly to changes in mechanical loads than does cortical bone. Cancellous bone has no Haversian system, whereas, cortical bone has Haversian system. Cancellous bone contains more osteoprogenitor cells, more metabolic activity so that fracture healing is faster in cancellous bone than cortical or dense bone. Similarly, remodelling is also faster in cancellous bone than cortical or dense bone.

Cortical or cancellous bone may consist of woven (fibre or primary) or lamellar (secondary) bone. Woven bone forms the embryonic skeleton or in fracture healing and then subsequently resorbed and replaced by mature bone (lamellar bone) as the skeleton develops. Woven bone are temporary or for a short period. In human, it disappears within the age of 4-5 yrs. Woven and lamellar bones differ with regard to their formation, composition, organization and mechanical properties. Woven bone has a rapid rate of deposition and turnover, whereas, lamellar bone is usually less active. Compared with lamellar bone, woven bone has an irregular, almost random, collagen-fibre pattern coarse (30μm in diameter) running in a random, interlacting or "burlap" fashion, and it contains approximately 4 times as many osteocytes per unit of volume. The osteocytes of woven bone vary in size, orientation, distribution, while those of lamellar bone are relatively uniform in size. The mineralization of woven bone follows an irregular pattern, mineral deposits vary in size. Collagen fibrils of lamellar bone very less in diameter (2-4μm), tightly organized and uniform in distribution. Woven bone is temporary whereas, lamellar bone is permanent. Woven bone is replaced by lamellar bone. There are four general forms of lamellar bone: (i) trabecular lamellae of cancellous bone; (ii) inner and outer circumferential lamellae of cortical bone; (iii) interstitial lamellae of cortical bone and (iv) lamellae of osteons. Each lamella consists of highly oriented, densely packed collagen fibrils.

Osteon: It forms the bulk of the diaphyseal cortex of mature skeleton. The central canals of osteons, referred as Haversian canals, containing blood vessels, lymphatic vessels, and occasionally nerves. Canaliculi containing the cell process of osteocytes extending a radial fashion from the central canal like the spokes of a wheel. These canaliculi connect the central canal to osteocytes and pass from osteocytes to osteocytes. The cells depend primarily on the canaliculi for the delivery of their metabolic requirements. Cement lines define the outer boundary of each osteon. This is the mark sites where the resorption of bone stopped and new bone formation began. The network of canals (Haversian + Volkmann, canaliculi and lacunae joint together) surface area is roughly100 times larger than the combined periosteal and endosteal surface area of mature cortical bone. The network of canals and lacunae forms an extensive extravascular space where ions and fluid can flow freely, directly adjacent to mineralized matrix.

Growth plate/epiphyseal plate: It is present in proximal and distal end of long bone. It's responsible for longitudinal bone growth and bone formation. The growth plate may be divided anatomically into three components: a) Cartilaginous component; b) Bony component/ Metaphysis and c) Fibrous component. Cartilaginous component has three zones: Reserve Zone, Proliferative zone and Hypertropic zone (have three sub zones: zone of maturation, zone of degeneration and zone of calcification). Fibrous component which present in the surrounding the periphery of the growth plate has two parts: groove of Ranvier and Pericordial ring of La Croix. Growth plate's closers depend upon the particular bone and breed and species. In Beagle dog, growth plate closure is (in days, average) of scapula-127, humerus-273, radius-136, ulna-204-236, carpus-101-110, metacarpus- 145-170, femur-129-136, tibia-200, fibula-136, tarsus-159, metacarpal-197 and phalanx—176.

Bone cells: Origin of bone cells: (i) Mesenchymal cell line, (b) Haematopoietic stem cell line.

Mesenchymal cell line (Pericytes) → Undifferentiated cells (Pre-osteoblast) → Osteoblast (active form) →Bone lining cells (resting form) →Osteocytes.

Haematopoietic stem cell line →Circulating/marrow monocytes→Pre-osteoclast→Osteoclast.

Function of bone cells: To carry out the diverse functions of the formation of bone, resorption of bone, mineral hemostasis and repair of bone.

Undifferentiated mesenchymal cells: Present in bone canals, endosteum, periosteum and bone marrow. They also may enter the bone by migrating form surrounding tissues or from the blood, vascular pericytes.

Function: Potentiality to convert pre-osteoblast and then osteoblast.

Structure: Irregular in size, a single nucleus, minimum cytoplasm and few organelles. They remain in their undifferentiated state until they are stimulates by growth factors or cytokines like TGF-beta or BMP to proliferate and differentiated into osteoblast.

Osteoblast: Present in osteoblast line in the surfaces of bone and pack tightly against adjacent osteoblast.

Structure: Rounded, oval, polyhedral form, single nucleus, cytoplasm contained ER, Golgi bodies, ribosome and mitochondria. The cytoplasmic process of osteoblast extended through the osteoid matrix to come into contact with osteocytes within mineralized matrix.

Synthesis: Collagen /organic matrix of bone.

Function: Formation of bone; bone lining; synthesis and secretion of organic matrix; controlling electrolyte fluids between the extracellular fluid and the osseous fluid; may influence the mineralization of bone matrix; osteoblast release mediators that active osteoclast under the stimulation of PTH and local cytokines; after its maturation surrounded by matrix, convert to osteocytes; actively play in bone repair by mitotic cell division → abundant osteoblast → abundant organic matrix synthesis → help in mineralization-bone repaired; it also attract osteoclast during absorption and remodelling of bone. Osteoblast has two forms: active form-take part in bone formation, in fracture healing and resting form-present in bone lining cell, elongated and flattened form, penetrate into the bone matrix by its cytoplasmic process and contact with osteocytes. Under stimulation, this resting osteoblast converts into active osteoblast.

Osteocytes: More than 90% of the bone cells in the mature animal skeleton are osteocytes. They surrounded themselves with an organic matrix that can mineralize.

Structure: Single nucleus, less cytoplasm, less organelles, long branching cytoplasmic process that extend into the surrounding matrix from some distance and fill most of the canaliculi in which they are contained. The cytoplasmic process of osteocytes, contact process from other osteocytes and osteoblast forming a large complex of network, tight junctions.

Function: Actively act in mineral ion movement through cytoplasmic process between bone matrix and extra vascular fluid spaces of the bone; provides compactness of the bone, it is very sensitive to any stress to bone.

Osteoclast: Origin: Unlike other bone cells, osteoclast appears to share a haematopoietic stem cell precursor with cells of Monocyte family. They are rarely found in normal bone, occasionally in Haversian and endosteal surfaces. In cancellous bone, they present in "How ship lacunae". Specific hormones and growth factors stimulate the stem cell to develop into osteoclast precursor.

Osteoclast precursor cells present in marrow and blood. When stimulated mononuclear osteoclast precursor cell → proliferate and fused into → large multinucleated osteoclast.

Structure: It has 3-22 nuclei, large number of mitochondria and lysozomes. Active form: Mitochondria fill much of their cytoplasm to supply the great amount of energy that is required to resorbed bone.

Distinct feature: Complex folding of cytoplasmic membrane (ruffled/brush boarder) where it is applied to the site of resorption of bone matrix → degradation of bone matrix.

Osteoclast and bone resorption: Osteoclast has an extremely efficient method for destroying bone matrix. They first bind themselves to the surfaces of the bone → creating a sealed space between the cell and the bone matrix → endosomes containing membrane bound proton pumps move to the region of the cell nearest the bone and insert into the cell membrane → the brush border of osteoclast active → proton pumps transport proton into the sealed space → decreasing the pH from approx 7.0 to 4.0 → the acid pH solubilize the bone mineral → to degrade the remaining organic matrix, the cell secreate acid phosphatese and acid proteases → in addition, the osteoclast may phgocytize some matrix fragements and degrade them within cytoplasmic vacules → then they move one site of resorption to another → after completion of resorption, they divide into mononuclear cells, which can be reactivated to form new osteoclasts.

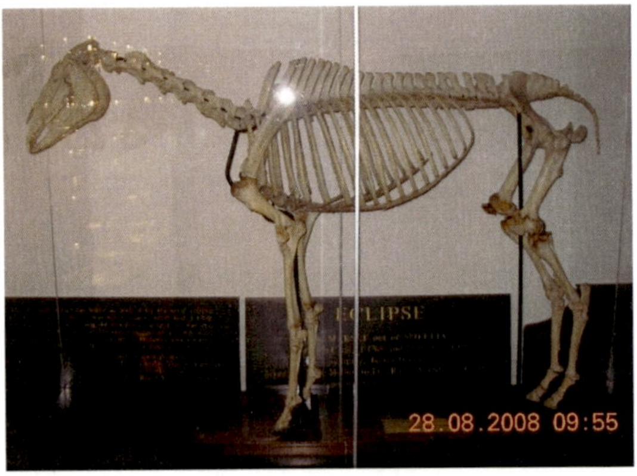

Horse "Eclipse" the cause for foundation of Royal Veterinary College, London

2

Fracture Types and Fracture Healing

A. Definition

A fracture is a break in the continuity of hard tissues like bone, cartilage, etc.

B. Etiology

(a) Extrinsic causes

- Direct violence
- Indirect violence

(b) Intrinsic causes

- Fractures due to muscular actions
- Fracture are common in certain bones because of their superficial position and shape
- Smooth road, slippery floors, sheds and stables
- Pathologic fractures: due to underlying bony or systemic diseases. eg: neoplasia, osteomylitis etc.

C. Classification of fractures

1. Based on presence of communicating external wound

i. **Closed fracture or simple fracture**: No communication with outside environment

ii. **Open fracture or compound fracture**: Fracture site communicating with external environment.

iii. **Complicated fracture**: A closed fracture in which there is considerable injury to important neighboring vessels or nerves or is accompanied by the opening of a joint or visceral cavity, e.g., complicated fracture of the rib.

2. A fracture may be incomplete or complete

(a) An incomplete fracture

It is a fracture which does not extend through complete thickness of the bone e.g., greenstick fracture, fissured fracture, deferred fracture, etc

i. **Greenstick fracture**- Fracture in which the bone is partly broken like bent green stick. Greenstick fracture occurs in young animals

ii. **Partial or splintered fracture**- When splinters of bone are separated from the main bone as a result of direct violence.

iii. **Fissured fractures:** There is a fissure extending from the bone without causing any displacement. It can longitudinal, transverse, and oblique. Seen in tibia, radius, metacarpus and os-suffraginis

iv. **Sub-periosteal (intra-periosteal) fracture**: A fracture of the cortical bone without rupture of the periosteum

v. **Deferred fracture**- It is an incomplete fracture in which separation of the fragments occur only after varying period.

(b) Complete fracture

Bone is broken completely through its thickness

i. **Single fracture**-Bone is broken in single place

ii. **Double fracture**-There are two fractures in the same bone

iii. **Multiple fractures**-When the bone is broken into more than two pieces.

3. Avulsion fractures

Tearing of bone prominences by forcible pull of its tendinous or muscular attachments

4. Based on the portion of the bone involved, fracture may be classified as:

i. **Diphysary fracture**- Fracture involving diaphysis of bone

ii. **Epiphysary fracture**-Fracture at the junction of the epiphysis and shaft of the bone.

iii. **Supracondylar fracture**-A fracture above the condyle e.g., supracondylar fracture of humerus

iv. **Condyloid fracture (condylar fracture)** - A fracture in which small fragments including the condyle is separated from the bone. Condyloid fracture of humerus, femur etc.

v. **Transcondylar fracture**-A fracture of humerus or femur in which the line of fracture is at the level of the condyles.

vi. **Intercondylar fractures**- A fracture between the condyles of the humerus. An intercondylar fracture and a transcondylar fracture may coexist, making a T-shaped fracture

vii. **Periarticular fracture**-When a bone is fractured close to its articulating extremity without extending into the joint.

viii. **Transcervical fracture**: Fracture through the neck of the femur

ix. **Articular fracture**- Fracture involving articular surface of the bone

x. **Extracapsular fracture**-A fracture near a joint but not entering within the joint capsule

xi. **Intracapsular fracture**-A fracture within the joint capsule.

5. Depending on the direction of the fracture, a fracture may be

i. **Transverse fracture**- A fracture at right angles to the axis of the bone

ii. **Longitudinal fracture**- A fracture extending in a longitudinal direction

iii. **Oblique fracture**- A break in the bone extending oblique in direction

iv. **Spiral fracture**- A fracture that is spiral in direction

6. Depending on the relationship between the fragments in the fracture

i. **Torsion fracture**- A fracture in which one of the fragments as been twisted and separated

ii. **Impacted fracture**—One fragment is firmly driven into another or one bone is driven into the fractured site of another

iii. **Dentate fracture**- A fracture in which the ends of the fragments are toothed and interlocked

iv. **Riding fracture (overriding fracture)** - A fracture in which the fragments lie side by side, causing shortening of the limb.

v. **Distracted fracture:** Fractures in which the fragments are separated by muscular pull, eg, fracture of olecranon.

7. **Salter Harris type fracture**

i. **Type I fracture** is when there is a fracture across the physis with no metaphysial or epiphyseal injury

ii. **Type II fracture** is when there is a fracture across the physis which extends into the metaphysis

iii. **Type III fracture** is when there is fracture through the physeal plate and epiphysis

iv. **Type IV fracture** is when there is a fracture through metaphysis, physis, and epiphysis

v. **Type V fracture** is when there is a crush injury to the physis.

8. A fracture also could be

1. **Compression fracture**-A fracture produced by compression, causing apparent reduction in the size of the bone due to pressure, e.g., vertebrae
2. **Depressed fracture**-A fracture of the skull, a fragment is depressed below the surface
3. **Collies fracture**- Fracture of the distal end of the radius.
4. **Monteggia fracture**- Is a fracture of the proximal third of the Ulna with dislocation of the proximal head of the Radius
5. **Pathological fracture**- A fracture occurring due to a weakening of bone by disease and not due to trauma
6. **Congenital (intra-uterine) fracture**: Facture of the bone of a fetus in the uterus

9. AO classification of fracture

The fracture of each long bone segment giving coding, according to their morphological characteristics are divided into *types*, *groups* and *subgroups*.

The Schematic diagram of the morphological characterization of the fractures. The colors green, orange and red indicates the increasing severity (A1-simplest fracture with best prognosis, C3-the most difficult with worst prognosis)

The 3 **Types** are labeled A, B, and C. Each type is divided into 3 groups: A1, A2 A3-B1, B2, B3-C1, C2, C3. Thus there are **9 groups**. Each group is further subdivided into 3 subgroups, denoted by a number 1, 2, 3. Thus there are for each segment 27 subgroups.

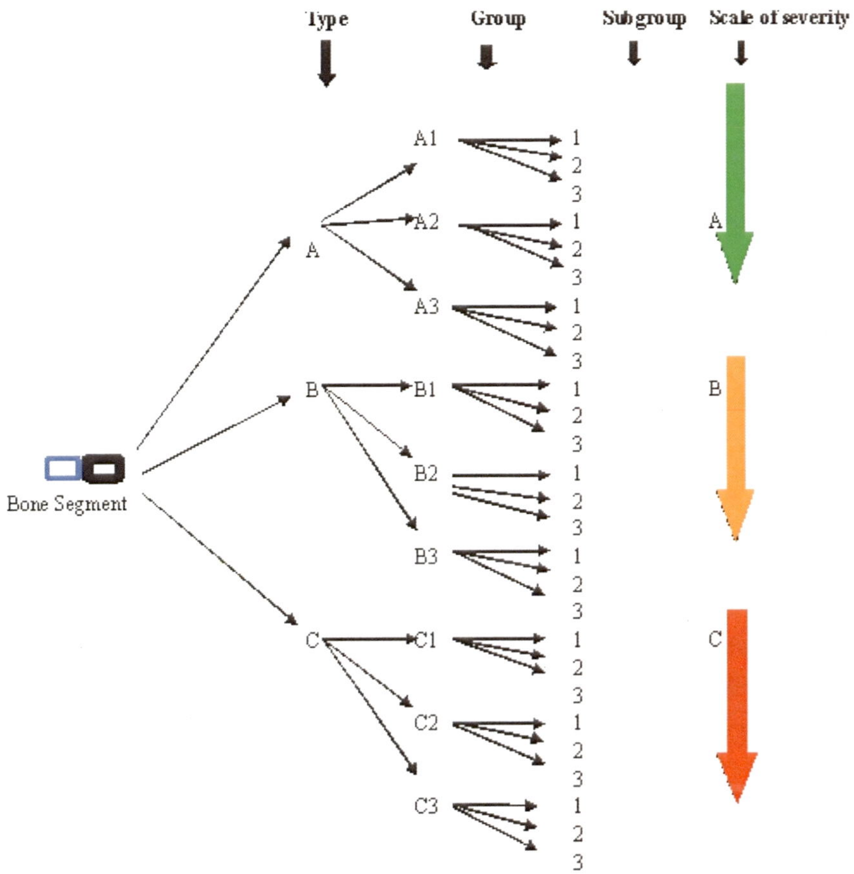

Terms used in AO Classification of fractures

All fractures are either **simple** or **multifragmentary**

Simple: A term is used to characterize a single circumferential disruption of a diaphysis or metaphysis or a single disruption of an articular surface. Simple fractures of the diaphysis or metaphysis are spiral, oblique or transverse.

Multifragmentary: A term used to characterize any fracture with one or more completely separated intermediate fragments. In the diaphyseal and metaphyseal segments, it includes the wedge and the complex fractures.

The term wedge and complex are used for diaphyseal or metaphyseal fractures.

- **Wedge**: A fracture with one or more intermediate fragments in which, after reduction, there is *some contact* between the main fragments. The spiral or bending wedge may be intact or fragmented.

- **Complex**: A fracture with one or more intermediate fragments in which, after reduction, there is *no contact* between the main proximal and distal fragments. The complex fractures are spiral, segmented or irregular.

  The term comminuted is imprecise and should not be used.
- **Impacted**: A stable and usually simple fracture of the metaphysis or epiphysis in which the fragments are driven into each other.
- **Extra-articular fractures:** These do not involve the articular surface although they may be intra-capsular. They include apophyseal and metaphyseal fractures.
- **Articular fractures**: Involve the articular surface. They are subdivided into partial and complete
- **Partial articular fracture**: The fractures involve only part of the articular surface while the rest of that surface remains attached to the diaphysis.
- **Pure split**: A fracture, resulting from a shearing force, in which the direction of the split is usually longitudinal.
- **Pure depression**: An articular fracture in which there is pure depression of the articular surface without a split. The depression may be central or peripheral.
- **Split depression:** A combination of a split and a depression, in which the joint fragments are usually separated.
- **Multifragmentary depression**: A fracture in which part of the joint is depressed and the fragments are completely separated.
- **Complete articular fracture**: The articular surface is disrupted and completely separated from the diaphysis. The severity of these fractures depends on whether their articular and metaphyseal components are simple or multifragmentary.

Coding of the diagnosis

To code the diagnosis of a fracture one must know its *location* and its *morphology*.

Location: this is designated by two numbers:

- Which bone?
  (1 humerus, 2 radius/ulna, 3 femurs, 4 tibia/fibula)
- Which segment?
  (1 proximal, 2 diaphyseal, 3 distal, 4 malleolar)

The proximal and distal segments of long bones are defined by a square whose sides are the same length as the widest part of the epiphysis

For coding the diagnosis, use an alpha-numeric code to conform to computer practice code.

Diagnosis Where?
Bone——————1, 2, 3, 4
Segment————1, 2, 3, 4
Diagnosis What?
Type——————A, B, C
Group—————1, 2, 3
Subgroups——————1, 2, 3

Example: Coding 32-B2.1 Means?

3	2	-B	2	.1
Femur	diaphysis	Wedge fracture	bending wedge	subtrochateric

Coding of different long bones

11-Humerus proximal

11-A extra-articular unifocal #
11A1 tuberosity
11-A2 impacted metaphyseal
11-A3 non impacted metaphyseal

11-B extra-articular bifocal fracture
11-B1with metaphyseal impaction
11-B2 without metaphyseal impaction
11-B3 with glenohumeral dislocation

11-C articular fracture
11-C1 with displacement
11-C2 marked displacement
11-C3with glenohumeral dislocation

12-Humerus Diaphysis

12-A Simple fracture
12-A1 spiral
12-A2 oblique (>30^0)
12-A3 transverse (<30^0)

12-B Wedge fracture
12-B1 spiral wedge
12-B2 bending wedge
12-B3 fragmented wedge

12-C complex fracture
12-C1 spiral
12-C2 segmental
12-C3 irregular

13-Humerus Distal

13-A extra -articular fracture
13-A1 apophyseal avulsion
13-A2 metaphyseal simple
13-A3 metaphyseal multifrag

13-B partial articular fracture
13-B1 sagital lateral condyle
13-B2 sagital medial condyle
13-B3 frontal

13-C complete articular fracture
13-C1 articular, metaphyseal simple
13-C2 metaphyseal multifragmentary
13-C3 articular multifragmentary

21-Radius/Ulna proximal

21-A extra articular fracture
21-A1 of ulna, radius intact
21-A2 of radius, ulna intact
21-A3 of both bones

21-B articular fracture
21-B1 of ulna, radius intact
21-B2 of radius, ulna intact
21-B3 of one bone, other extra-articular

21-C articular fracture of both bone
21-C1 articular simple
21-C2 one bone simple & another frag
21-C3 articular multifragmentary

22-Radius-Ulna diaphysis

22-A Simple fracture
22-A1 of ulna, radius intact
22-A2 of radius, ulna intact
22-A3 of both bones

22-B wedge fracture
22-B1 of ulna, radius intact
22-B2 of radius, ulna intact
22-B3 of one wedge, other simple/wedge

22-C complex fracture
22-C1 of ulna
22-C2 of radius
22-C3 of both bones

23-Radius/Ulna distal

23-A extra-articular fracture
23-A1 of ulna, radius intact
23-A2 of radius simple, impacted
23-A3 of radius, multifragmentary

23-B partial articular fracture
23-B1 of radius, sagital
23-B2 of radius, dorsal rim
23-B3 of radius, frontal, volar rim

23-C complete articular # of radius
23-C1 articular, metaphyseal simple
23-C2 metaphyseal multifragmentary
23-C3 articular multifragmentary

31-Femur proximal

31-A extra-articular #, trochanteric
31-A1 protrochanteric simple
31-A2 protrochanteric multifragmentary
31-A3 intertrochanteric

31-B extra-articular #, neck
31-B1 subcapital, slight displacement
31-B2 transcervical
31-B3 subcapital, displaced

31-C articular #, head
31-C1 split (Pipkin)
31-C2 with depression
31-C3 with neck fracture

32-Femur diaphysis

32-A simple fracture
32-A1 spiral
32-A2 oblique (>30⁰)
32-A3 transverse (<30⁰)

32-B wedge fracture
32-B1 spiral wedge
32-B2 bending wedge
32-B3 fragmented wedge

32-C complex fracture
32-C1 spiral
32-C2 segmental
32-C3 irregular

33-Femur distal

33-A extra-articular #
33-A1 simple
33-A2 metaphyseal wedge
33-A3 metaphyseal complex

33-B partial-articular #
33-B1 lateral condyle, sagital
33-B2 medial condyle
33-B3 frontal

33-C complete articular #
33-C1 articular, metaphyseal simple
33-C2 metaphyseal fragmentary
33-C3 articular fragmentary

41-Tibia/Fibula Proximal

41-A extra-articular #
41-A1 avulsion
41-A2 metaphyseal simple
41-A3 metaphyseal multifragmentary

41-B partial articular #
41-B1 pure spilt
41-B2 pure depression
42-B3 split-depression

41-C complete articular #
41-C1 articular, metaphyseal simple
41-C2 metaphyseal multifragmentary
41-C3 articular multifragmentary

42-Tibia/Fibula diaphysis

42-A simple fracture
42-A1 spiral
42-A2 oblique (>30⁰)
42-A3 transverse (<30⁰)

42-B wedge fracture
42-B1 spiral wedge
42-B2 bending wedge
42-B3 fragmented wedge

42-C complex fracture
42-C1 spiral
42-C2 segmental
42-C3 irregular

43-Tibia/Fibula distal

43-A extra- articular #
43-A1metaphyseal simple

43-B partial articular #
43-B1 pure spilt

43-A2 metaphyseal wedge
43-A3 metaphyseal complex

43-B2 split-depression
43-B3 multifragmentary depression

43-C complete articular #
43-C1 articular, metaphyseal simple
43-C2 metaphyseal multifragmentary
43-C3 articular multifragmentary

44-TibiaFibula Malleoli

44-A infrasyndesmotic lesion
44-A1 isolated
44-A2 medial malleolus #
44-A3 with postero-medial #

44-B transsyndesmotic fibular #
44-B1 isolated
44-B2 with medial lesion
44-B3 with medial lesion + Volkmann

44-C suprasyndesmotic lesion
44-C1 fibular diaphyseal # simple
44-C2 -do- multifragmentary
44-C3 proximal fibular lesion

D. Signs of Fracture

- Deformity-shortening/angulations/ rotation/adduction/abduction
- Loss of function-inability to bear weight
- Pain
- Local trauma
- Abnormal posture or limb positioning
- Crepitus
- Abnormal mobility
- Radiographic signs
- Other signs: Rise of temperature, Albuminuria, Lipuria

E. Diagnosis of Fracture

- History
- Physical Examination
- Clinical Signs
- Radiography (X-Ray)
- Computed Tomography (CT)
- Magnetic Resonance Imaging (MRI)
- Bone Scan

F. Healing of the Fracture

- Primary healing: Bone is the first and only type of connective tissue formed
- Secondary healing: Formation of cartilaginous or fibrous tissue first, subsequently become mineralized & replaced by bone. Indirect (secondary)

fracture healing is the most common form of fracture healing, and consists of both endochondral and intramembranous bone healing. It does not require anatomical reduction or rigidly stable conditions.

Primary healing

- Contact healing: Bone on one side of the cortex must unite with bone on the other side of the cortex to re-establish mechanical continuity. If the gap between bone ends is less than 0.01 mm and interfragmentary strain is less than 2%, the fracture unite by "contact healing". Cutter cones are formed at the ends of the osteons closest to the fracture site. The tips of the cutting cones consist of osteoclasts which cross the fracture line, generating longitudinal cavities. These cavities are later filled by newly developed osteoblasts. These results in the simultaneous generation of a bony union and the restoration of Haversian systems formed in an axial direction. The bridging osteons later mature by direct remodelling into lamellar bone resulting in fracture healing without the formation of periosteal callus.
- Gap healing: It is differs from contact healing in that bony union and Haversian remodelling does not occur simultaneously. It occurs if stable conditions and an anatomical reduction are achieved, although the gap must be less than 800 μm to 1mm. In this process the fracture site is primarily filled by lamellar bone oriented perpendicular to the long axis, requiring a secondary osteonal reconstruction unlike the process of contact healing. Vascular loop accompanying osteoprogenitor cells grows into fracture gap and differentiate into osteoblasts.

Characteristics of primary healing

1. Anatomical reduction, motionless stability & interfragmentary compression favors primary healing
2. Periosteal or endosteal callus is absent
3. Cellular function & vascularization proceeds unimpaired
4. Resorption & replacement of fracture ends does not occur.
5. Does not occur sooner than secondary union.
6. The bone undergoing primary union is mechanically weaker than secondary union.
7. Use of rigid fixation lead to decreased cortical density in early stages of fracture fixation

Secondary healing

It is as also called classical healing. It comprises of three phases: inflammatory phase, reparative phase and remodeling phase.

a. Inflammatory phase

i. Extensive hemorrhage

ii. Ischemic necrosis of bone

iii. Fibrin clot formation

iv. Increase in extra-osseous blood supply

v. Sprouting of new blood vessels & proliferation of fibroblasts

b. Reparative phase

i. External or periosteal callus and internal or endosteal callus formation.

ii. Derived from pluripotent mesenchymal stem cells, cambium layer of periosteum, & marrow cavity.

iii. Early stage: fibrous and cartilage predominates depending on regional microenvironment: sub stage of soft callus

iv. Later stage: mineralization: sub stage of hard callus.

v. Cartilage is reabsorbed in a process similar to endochondral ossification

vi. Necrotic bone reabsorbed by osteoclasts

c. Remodeling phase

i. Slow change in the shape of the fracture & bone.

ii. Governed by Wolff's law

iii. Eventually, the compact bone of cortex is restored to its normal osteonal architecture

G. Factors affecting fracture healing

a. Favorable factors

1. Rigid immobilization.
2. Fracture at the ends of the bone, in areas of cancellous bone & adequate blood supply.
3. Minimal surrounding soft tissue damage.
4. Fracture site free of infection.
5. Younger animals heal faster than older animals

b. Unfavorable factors:

- Local factors: eg. Degree of local trauma, degree of loss of bone, type of bone, local malignancy, infection

- Systemic factors: age, hormones, corticosteroids, hypo-proteinemia, diabetes etc.

H. Fracture Complications

- Delayed union: It occurs when a fracture fails to heal within the expected time, in old animal and in patients where there is limited movement of the fracture site. Inadequate reduction and fixation, poor vascularity at the fracture site, extensive periosteal and surrounding soft tissue damage, metabolic diseases, and mineral, vitamin and protein deficiency are the other causes of delayed union. It is generally found in long bones likes radius and ulna, tibia and femur. Diagnosis can easily be done by clinical and radiological examination. Increased stabilization at the fracture site is essential in this condition.
- Non-union: Is the most serious form of fracture complication due to many causes. Biological inactivity of the fracture is a less common cause than movement at the fracture site. The imposition of soft tissues, dead bone, infection, dietary deficiency of mineral, vitamin and protein, and excessive activity may all predispose to this condition. Bone drilling at the sclerosed fracture ends, bone grafting, and increase bone fracture ends stabilization either by bone plating or by external fixation, pulsed or continuous electrical stimulation either by invasive or semi-invasive method and prolonged physiotherapy is recommended for treating of non-union.
- Malunion: It results either from a failure of the initial reduction or from displacement of the reduction followed by secondary callus formation. In this condition, fracture ends unite in an abnormal position. Improper internal or external fixation techniques, fixation failure, inadequate fracture fixation, improper positioning of the animal during external fixation, excessive movement during post fixation are the causes for malunion. Diagnosis can be done by clinical and radiological examination. Malunion can be treated by secondary operations-osteotomies of the involved bone and then fixation by rigid fixation technique.
- Soft tissue abnormalities: Are frequently induced by surgeons during fracture management. Infection and careless surgery can result in adhesions, muscular fibrosis, joint stiffness and neurological damage. Excessive soft tissue damage during internal fixation leads to less blood /nerve supply to the fractured bone ultimately causes delayed or non-union.
- Bone atrophy
- Osteoarthrosis: Degenerative joint disease may be a sequel to a fracture involving an articular surface.

- Osteomylitis: It may be defined as inflammation of bone and bone marrow. Exogenous and/ or haematogenous bacterial infection is the main cause of osteomylitis. Bacteria may invade during fracture (compound fracture), during operation or during post fixation/post fracture. Long bones are generally involved. Diagnosis by clinical signs (lameness, pus formation, bad odour, swelling at the fracture site); radiological signs (periosteal reaction, cortical lysis, sunburst appearance), microbial examination of pus or exodus and bone biopsy. Heavy and prolonged antibiotic therapy as per antibiotic sensitive test, daily drainage of pus and wound dressing, obliteration of dead space, wound debridement, removal of necrosed bone pieces, bone grafting and stable fixation are recommended for treatment of osteomylitis.
- Fracture-associated neoplasia: Fracture-associated sarcoma is a rare fracture complication.

3

Role of Different Minerals Vitamins and Hormones in Fracture Healing in Animals

Fracture healing is a specialized type of a wound healing response in which the regeneration of bone leads to a restoration of skeletal integrity. Because healing of a fracture in a living organism is accomplished by cells, it can be modified by almost any endogenous or exogenous factors like nutrients, vitamins, hormones, growth factor etc. that has an influence on the metabolic function of bone cells. Fracture healing may be induced by malnutrition that involves certain nutrients, notably calcium (Ca), phosphorus (P) and vitamin D, but also iodine (I), copper (Cu), zinc (Zn), manganese (Mn), fluorine (F), silicon, vitamin A and perhaps vitamin C as well as different endocrines. The number of these nutrients and their interaction is so large that dietary intervention in fracture management is likely to be the most effective if directed at consumption of a good diet that is completely balanced for growth, maturity, old age or stress. Substitution of a good diet for a bad one is also likely to be more convenient and safer than most attempts at partial supplementation of a faulty diet in order to bring it into balance. Underfeeding a good diet decreases size-for-age of bones during growth and delayed fracture healing. Underfeeding may engender osteoporosis in older animals. Underfeeding of Ca, P and vitamin D, causes delayed fracture healing. Over feeding a balanced diet induces rapid growth of newly formed bone and sometimes-accelerated fracture healing but it also causes obesity in adults.

Three major component of bone are osteogenic cells, organic matrix and inorganic matrix (mineral). The osteogenic cells include osteoblasts, osteocytes and osteoclasts, while the matrix consists predominantly of collagen and proteoglycans and constitutes approximately one-third of bone mass (on dry basis). The inorganic/mineral component that makes up approximately two-third of bone and it performs two essential functions: it serves as an ion reservoir and it gives bone most of its stiffness and strength. Bone is a reservoir for 99% of body calcium, 85% of body phosphorus and 40-60% of body sodium and magnesium. The main mineral composition of a long bone is: 85% calcium phosphate as hydroxyapatite, 12% calcium carbonate and 3% magnesium phosphate, calcium fluoride and magnesium carbonate and trace amount of chloride, potassium and sodium. For collagen synthesis, hydroxylase enzymes requires ferrous ion and vitamin C. Impaired collagen synthesis also occurs during deficiencies of vitamin D and E and zinc. The organic intercellular bone matrix is rich in mucoplysaccharide. Their synthesis involves enzymes that require manganese, zinc and vitamin A.

Calcium is the most important mineral constituent of the skeleton. It is also well established that calcium supplementation accelerate fracture healing by early calcification of newly formed woven bone. For Ca absorption in the gut, phosphorus along with vitamin D is essential. In the Ca-deficient animal i.e. rickets, osteomalacia, osteoporosis, fracture healing either delayed or causes non-union. Under experimental condition, a diet deficient in calcium, phosphorus and vitamin D induced delayed fracture healing in animals-leading to rickets and complicated by osteoporosis; particularly those are high-yielding animal. Supplemented with vitamin D reduced the rickets but not osteoporosis. Excess dietary calcium also leads to abnormalities in fracture healing and bone development. It is suggested that excess calcium impaired the absorption of copper and hence induced osteoporosis and also interfered with thyroid uptake of iodine, causing hypothyroidism and the short bone and delayed healing. Finally, the dietary Ca: P value has been shown to have a major effect on the rate of fracture healing. Fracture healing abnormalities occurs either by decreasing the dietary Ca content as constant P content or by increasing the P content at adequate Ca content.

The mineral phosphorus (P) is as important for bone development during fracture healing, as Ca, since it makes up about half the weight of bone mineral and hence, must be available in adequate amounts in the daily animal diet for proper mineralization of bone/callus. It is present in relatively adequate amounts in the daily diet and attention, therefore, has tended to focus on the potentiality harmful effects of excessive amounts and characteristics of a low Ca: P value. Increased phosphorus intake has been shown to depress ionized Ca. However, there is some evidence to show that increased P intake reduces Ca loss; increasing Ca absorption and thus total body Ca is not affected.

Recently, attention has been focused on the influence of potassium (K) on bone health and bone formation. The skeleton play an important role in acid-base homeostasis by the metabolization of skeletal salts to balance the endogenous acid generated from acid producing foods. The skeleton is not only a liable reservoir of K responsive to the mechanism, which maintains ionic Ca, but it also contains a reservoir alkaline salts Ca, which in turn provides a source of labile base. This can be mobilized to reacts both blood pH and plasma bicarbonate concentrations. The failure to maintain adequate acid base homeostasis in adults has been proposed as the main reason for the progressive decline in bone mass, which with age. Such bone loss may be due to the lifelong mobilization of skeletal Ca salts to balance the endogenous acid generated form food such as meat.

There is much evidence supporting a positive relationship between urinary sodium and Ca excretion in adult animals consuming a normal diet. Re-absorption of Ca and Na in the proximal tubule is known to be linked. Thus, a reduction in renal Na re-absorption (such as that induced by a high salt diet) also leads to a reduction in Ca absorption and increased Ca urinary loss. Hypercalciuria, caused by high Na intakes is well established in both human and animals.

The nutrient magnesium (Mg) is considered by many to be extremely important for skeletal development. In animal studies, Mg deficiency causes cessation of bone growth and decreased osteoblastic and osteoclastic activity, osteopenia, increased bone fragility and the development of a form of "aplastic bone disease".

Zn, Cu and Mn are known to be essential metallic co-factors for enzymes involved in the synthesis of various bone matrix constituents during new bone formation followed fracture. Mn is essential for the development of organic matrix of bone, which is largely composed of mucoplysaccharide, and impairment in mucoplysaccharide synthesis is largely associated with Mn deficiency. Mn is concerned with the activity of glycosyl transferase. Ataxia and impaired bone cell formation have been linked to glycosaminoglycans metabolism. In Cu-deficient animal, fracture healing is generally delayed. Skeletal abnormalities occur in almost all species of animal and poor mineralization of new bones responsive to Cu- supplementation also occurs. Positive association between dietary intake of Zinc (Zn) and new bone development has been cited in different literature. Skeletal abnormalities are predominant feature of Zn deficient animals likes shortened and thickened legs, dorsal curvature of spine etc. Leg bone abnormalities in Zn deficient animals are reversible on Zn feeding. In Zn deficient animal, fracture-healing retarded /delayed may be due to decreased osteoblastic activity and reduction in chondrogenesis.

Different vitamins also have a great role in the management of fracture. Vitamin D and K are two fat-soluble vitamins that play different but essential roles in fracture healing particularly in new bone mineralization. Vitamin D, after its conversion to a dihydroxylated metabolite, acted as a classical steroidal hormone that regulates the transcription of manifold target genes involved with Ca homeostasis in general and bone mineralization in particular. It is well established that in vitamin D deficient animal, fracture healing either delayed or non-union, while vitamin D supplemented animal fracture healing is accelerated. The role of vitamin D in promoting mineralization of newly formed woven bone during fracture healing reparative process has of course been appreciated for many years, together with the deficiency syndromes of rickets and osteomalacia that results from prolonged deficiency states in young and adulthood, respectively. Vitamin D [1, 25 $(OH)_2$ D] is essential for Ca absorption from the gut and Ca mobilization in the bone.

Vitamin K, on the other hand, is a co-factor for a carboxylase enzyme that resides in several tissues, also play a positive role in fracture healing. Vitamin K is essential for synthesis of several vitamin K- dependent bone proteins like osteocalcin or Gla protein (BGA), which accounts up to 15% of the non-collagenous bone protein. Besides, osteocalcin, vitamin K is also essential for synthesis of other two bone proteins namely matrix Gla protein (MGP) and protein S. The function of these bone proteins is unclear.

Other vitamins like vitamins A, B and C also have role in fracture healing. Vitamin A deficiency in feed in animal causes retardation of skeletal growth and bone remodelling and maturation. Vitamin B deficiency, particularly riboflavin, may produce congenital deformities of skeletal bone. Vitamin C is necessary for the production of intercellular cement substances during callus development of reparative phase of fracture healing. Therefore, a deficiency results in a reduced capacity of osteoblast to form osteoid. Failure to produce osteoid interferes with endochondral ossification and thus produce delayed fracture healing. The bone that does form, a slender and fragile during fracture healing. Blood vessels, particularly capillaries are fragile and easily ruptured, since multiple capillary hemorrhages are feature of vitamin C deficiency, during bone development. Deficiency of vitamin E may be related to bone growth disturbances particularly it inhibits the formation of collagenous connective tissue.

Like minerals, vitamins, different endocrines also have a great role in fracture healing. Some have acceleratory effect on fracture healing while some causes retard fracture healing.

The main function of the growth hormones seems to be that it directly stimulates the skeleton growth and its action in animal depends upon the fact whether epiphyseal growth plates of the long bones have ossified or not. If they have not at the time of over production of this hormone, gigantism results and if they united, the terminal part of the bones grow further leading to the development of acromegaly. The experimental studies to find out the mode of action of the hormone revealed that it produces nitrogen balance and raises total protein level. Growth hormone, therefore, seems to be a metabolic hormone presenting remarkable anabolic activity. Its own only drawback is that it mobilizes sugar from the liver into general circulation leading to high blood sugar level if excess of growth hormone is given. In different fracture model study it showed accelerated rate of fracture healing. In these experiments, the animal receiving growth hormone alone showed increased deposition of collagen fibers in the first week of post-fracture, leading to quicker and direct transformation of collagen fibers into bone without having the intermediary process of osteochondral ossification. At the end of 4th week of post-fracture, good amount of new bone formation in between the fracture fragment to bridge the gap could be seen. Thus, somatotropic (STH) or growth hormone (GH) exerts an accelerating and strengthens effect on the maturation of newly formed woven bone during fracture healing process.

Thyroid hormones are essential for skeletal maturation, the formation and development of epiphyseal centers in the long bones and growth of flat bones. Thyroid hormones have a direct effect on bone maturation *in vitro*. It is stated that action of thyroid hormone on the skeletal system is complex and influences all phases of endochondral osteogenesis. It has been suggested that thyroid hormone increase the conversion of the labile bone calcium into the more stable form and reduce turnover the later. In experimental studies, administration of thyroid hormone/thyroxine in physiological doses accelerated certain phases of bone healing process without any untoward effects on body. It is also observed that administration of thyroxine along with growth hormone have synergistic action on each other so far as the regeneration of new bone is concerned during fracture healing process.

Calcitonine, a polypeptide synthesized by the C cells of the thyroid gland, lowers the level of serum calcium by inhibiting osteoclastic resorption of bone. This may occur by inhibition of osteoclast precursor proliferation or inhibition of the formation and activation of osteoclast. The effect of calcitonine appears to be short term, thus its role in the physiology of normal bone is unclear. The hypocalcaemic effects of CT are primarily the result of decreased entry of calcium from skeleton into plasma owing to a temporary inhibition of PTH-stimulated bone resorption. Because of its potentiality to decrease the resorption of bone, it is experimentally used in the treatment of non-union, delayed union and on different fracture models. In general, Tri-iodothyronine (T3), thyroxine (T4), thyrotropin (TSH) and calcitonine accelerated fracture-healing process.

Parathyroid hormone (PTH) is the principal hormone involved in the minute-to-minute fine regulation of blood calcium in animals. The action of PTH on bone is to mobilize calcium from skeletal reserved into extracellular fluids. The role of PTH in the development, growth and repair of bone, particularly their effect on the process of calcification of the newly formed matrix, is a comparatively less investigated. This hormone is found to exert their effect more on the organic constituents than on the actual calcification process. PTH causes lesser collection of collagen fibers with formation of cartilage. This excess of cartilage slowly underwent the process of endochondral ossification and new bone formation. The new bone formed in this way is thinly trabeculae and found to be weaker in strength. So, it can be concluded that PTH delayed fracture-healing process by inhibiting the synthesis of collagen during reparative stages of fracture healing.

In view of it anabolic and growth promoting properties, insulin has direct role in fracture repair and regeneration of new bone. Physiologically, insulin is associated with proper utilization of glucose and hexosephosphates, the later ones being important source of phosphorus for calcification of newly developed bone matrix. It has been found that administration of insulin in physiological doses not only accelerated fracture healing but also soft tissue wound healing. On the other hand, it has also been reported that the insulin deficiency delayed fracture as well as wound healing. Insulin promotes the synthesis of bone organic matrix – mucoplysaccharide and induces better pattern of collagenization and causes earlier ossification/fracture healing. All this is due to anabolic effect of the hormone on the osseous tissue.

The action of sex hormone is not confined to the growth and development of the sexual organs alone, but the normal body growth as well as the growth of skeleton is to a great extent influenced by these hormones. *In vitro* studies on fracture healing has shown that the low concentration of testosterone enhanced healing, while higher concentration delayed/retard fracture healing. It is well documented that following oopherectomy, or in the post menopausal, females experiences a "rapid phase" of bone loss for the next few years. It has been demonstrated that continued administration of estrogens or progesterone to oophorectomised females would delay the onset of bone loss. In our study, progesterone supplementation in smaller doses, enhanced fracture healing by markedly increase osteoblast population as evinced by radiological and histopathological observations. However, in high dose, progesterone like testosterone and estrogen, retard fracture healing. Castration also has shown to retard fracture healing in experimental situations.

In both experimental and clinical situation, corticosteroids are powerful inhibitors of the rate of fracture healing. They have shown to inhibit the differentiation of osteoblast from mesenchymal cells and to decrease the rate of synthesis of the major components of bone matrix, which are necessary for fracture repair.

Prostaglandin (PG), a family of cycled derivation of unsaturated long chain fatty acids, has been found to be intimately associated with bone metabolism. It has been observed that long-term intravenous therapy with PGE1 to maintain circulation of infants with cardiac defects leading to periosteal proliferation on the surfaces of long bones. PGE2 given orally for 90 days has also been shown to produce increase response in periosteal bone formation. When prostaglandin (PGE1and PGE2) injected locally in small doses, resulting in more periosteal and cortical bone formation, resulting in accelerated fracture healing. In contrast, higher concentration of prostaglandin suppressed the cellular activities of bone cells.

4

Bone Grafts, Bone Substitutes and Bone Bioimplants in Small Animals Orthopaedics

1. Bone grafts

Bone grafting represents one of the earliest devised reconstructive approaches to the musculoskeletal system and remains among the most commonly used orthopedic procedures. Bone is the second most frequently transplanted tissue. The first recorded attempt to use bone graft is by the Dutch surgeon Job Van Meek'ren in 1668.

Definition

A graft is a living tissue that is transplanted to another site either within the same individual or into another individual for repair of a defect. An implant is a non viable material placed in a living system. Implants are composed of bone cement, metal or ceramic material. Grafts are structurally important for mechanical stability, but their use evolved to include the stimulation and enhancement of bone in response to injury.

Indications

Bone grafts are used, in general as framework, both to provide stability and to augment healing. It is used in the management of non union, delayed union, treatment of pseudoarthrosis, filling of osseous defects or cavities, arthrodesis, stabilization of spinal segments and addition of bone stock in total joint displacement.

Functions of bone grafts

Depending on the type of graft, bone performs four different functions when it is incorporated into host tissue .These include osteogenesis, osteoinduction, osteoconduction and mechanical support.

i. Osteogenesis

It is the formation of new bone initiated by donor cells such as osteoblasts and osteoclasts that survive transfer to a recipient site. This process that occurs in fresh cancellous autograft because they can revascularize more rapidly than fresh cortical autograft, allowing for survival of transferred cells. Also, cancellous bone has large surface area covered by quiescent lining cells, undifferentiated mesenchymal cells or active osteoblasts therefore this tissue has a greater potential to stimulate the production of new bone than other types of bone grafts. Osteogenesis can be initiated as early as 5 days after implantation and is maximized by 8 weeks following transplantation. Both autogenous and allogenic cancellous bone grafts are completely incorporated into the host bone by one year.

ii. Osteoinduction

It is the differentiation of undifferentiated mesenchymal cells into osteogenic cells (osteoblasts) after coming into contact with bone matrix. The major phases of this process are chemotaxis, mitosis and differentiation. Mesenchymal cells are stimulated to become chondroblasts and osteoblasts that proliferate and produce mineralized matrix. The differentiation is modulated by many factors including bone morphogenetic proteins (BMPs), transforming growth factor beta (TGF-â), insulin-like growth factor, cytokines, platelet derived growth factor, tumor necrosis factor and prostaglandin E_2. Both cancellous and cortical bones contain BMPs within its matrix, but cancellous bone has a larger percentage of matrix proteins exposed (primarily BMPs) to the cellular environment than cortical bone as a result of the large surface area associated with trabecular architecture.

iii. Osteoconduction

It is the passive ingrowths of new bone by extension from recipient bed. It is adheres dimensional process of in-growth of capillaries perivascular tissue and mesenchymal cells from the recipient bed into the structure of an implant or graft. The graft provides scaffolding for this growth. The term creeping substitution has been used to describe this dynamic, reconstructive and healing process of bone transplantation, where new viable bone replaces the old necrotic one. This nomenclature doesn't accurately describes the osteogenic response to autogenous cancellous bone graft in a recipient site, however, instead new bone is applied directly over existing necrotic trabeculae, with subsequent resorption of the necrotic trabeculae occurring following the deposition of new bone.

iv. Mechanical support

Grafts that provide mechanical support is weight bearing struts that fill in large bone defects, therefore providing strength and structure. Typically, these are whole allogenous cortical sections. To ensure stability of the graft within the recipient site, the graft must be stabilized with compression developed at the host –graft interface by using rigid internal fixation e.g. dynamic compression plating.

Although bone grafts facilitate healing and provide support through the above methods, there are also disadvantages to this form of bone grafting. It is important to evaluate the positive and negative effects when deciding whether to use a bone graft and the type of bone graft to be use. Disadvantages of bone grafting depends on the type of graft used and include (1) surgical invasion of the donor site and limited bone source in autograft (2) immune-mediated rejections of allograft and all implants (3) transmission of infectious agents with allograft and alloimplant (4) technical difficulty with vascularized grafts (5) fracture pain, seroma, hemorrhage and infection at the harvesting sites (6) infection and instability at graft site that results in failure

Classification

Bone grafts are classified according to their origin, the type of bone used (structure), or the graft placement technique:

i. By origin

a) Autograft –transplanted material from the same individual

b) Allograft-transplanted material from the same species of animals

c) Xenograft-transplanted material from different species of animals

d) Isograft-when tissue from one twin is implanted into an identical, monozygotic twin.

ii. By structure

a) Cortical graft: Harvested from the cortical part of the bone. A cortical bone graft can produce a good mechanical filling of a defect and the desired functional results, even though it may take much longer time to become viable. Cortical bone grafts have strictly limited clinical applications. Since they are primarily used in an area where there is great mechanical stress, proper fixation has to be used to allow these grafts to produce function needed. In the cortical autograft, the main differences with the cancellous revolve around the amount of revascularization and completeness of the remodeling. The cortical bone may not be vascularized as quickly as the cancellous graft. Revascularization takes about 2 months due to the structure of the cortical graft, which does not allow a large contact area for vascular penetration between the graft and the host.

b) Cancellous graft: Harvested from the cancellous part of the bone. In the majority of cases, cancellous bone is used because of its extensive ease of application for achieving fusion and for connecting discontinuity defects. It can also be used in clean or grossly contaminated wounds. It is generally harvested from the wing of the ilium, the proximal tibia and the proximal humerus. Cancellous bones usually do not have the mechanical strength desired for application in large defects. Therefore rigid fixation bridges the discontinuity defect area to alleviate the mechanical stress or is used to fix a graft in place when it is used as an onlay or for fusion. Because of the large open areas in these grafts, revascularization usually takes place very well, thereby bringing new cellular regeneration, remodeling and substitution, with new bone formation as the old bone is removed. The entire cancellous bed may be completely vascularized within approximately 1-2 weeks. The cell population of this environment comprises predominantly osteoblasts. These grafts can be applied in open fractures or in oral cavity defects, which are considered clean contaminated wounds. A cancellous bone graft can produce the desired filling and healing much better and faster than a graft of cortical bone in a contaminated area, which may end up being sequestered and have to be removed.

c) Cortico cancellous graft- Harvested from both the cortical and cancellous parts of the bone. Cortico cancellous bone grafts usually produce the best results because it enables good vascularisation to help the incorporation of the bone graft with the surrounding structures and also gives good mechanical strength. A rigid fixation apparatus can be used to produce the desired contour, particularly in a discontinuity defect application.

d) Osteochondral graft: Composed of bone and articular cartilage

e) Particulate bone-Composed of small chips of bone and is usually applied in an area where there is no need of mechanical strength. This is used for discontinuity defects where the defect is also bridged with a mechanical device, internal or external, and the defect is filled with these particulate bones.

f) Bone slurry: It is a mixture of small particulate of bone, usually realm of 100-250 microns. These are usually ground from cortical or cortico-cancellous bone and mixed with blood and other components, such as microfibular collagen, to give them form for application in certain defects. Bone slurry grafting is used to achieve quick revascularization and also to treat non stress –bearing areas. The area also has to be protected during the healing process, so either an external or an internal appliance is used.

g) Bone paste: Small fragments of bone of different sizes compact together, which is eventually, are incorporated in the surrounding structure. It is usually similar to slurry in that components are in a mixture that appears gelatinous and does not have structure, but its biologic nature allows quick revascularization of the graft so that quick healing, remodeling, solidification and incorporation can take place.

h) Composite graft: Fresh cancellous bone graft mixed with autogenous bone marrow or autogenous serum.

iii. By grafting technique

a) Onlay graft: It consists of primarily cortical implants placed along existing bone, thus bridging an osseous defect. Single or dual onlay grafts are placed in the recipient bone, which is previously drilled with small holes, and then fixed by compression screws. Cancellous bone fragments are often added to stimulate osteogenesis. Compression plates in combination with cancellous bone graft have largely replaced these methods.

b) Inlay graft: A rectangular segment of cortical bone is removed and then inserted into a recess created across the bone defect. Inlay grafts are originally advocated for treatment of fractures and non-unions. The sliding inlay graft consists of a longer rectangular section of bone from the proximal fragment and a shorter rectangular segment from the distal fragment. The distal fragment is removed. Sliding it distally to bridge the osseous defect displaces the longer proximal section. The shorter fragment is then inserted into the defect created by sliding the longer fragment distally. At present, there is only limited application of this technique in achieving arthrodesis or in supplementing internal fixation with compression plates.

c) Dowel grafts: This type of bone grafts has been used in the treatment of Navicular fractures, depressed tibial plateau fractures and non-united femoral neck fractures. A round core of cancellous bone is inserted into a surgically created channel across a fracture or non-union site. The cancellous bone promotes osteogenesis.

d) Muscle pedicle-graft: This has been used primarily for femoral neck fractures and non-union, particularly in young patients, and has largely replaced the use of dowel grafts in the site. These types of bone grafts have also been used in the treatment of femoral head osteonecrosis. The muscle pedicle bone graft is removed from the posterolateral aspect of intertrochanteric crest with the quadrates femoris muscle insertion. The bone graft with muscle insertion is then transplanted into a surgically created slot in the posterior femoral neck, transfixing the fracture site. A single screw secures the bone graft at each end with an additional internal fixation device. It appears that a significant amount of osseous graft survives with the muscle pedicle technique.

e) Strut graft: These are commonly used in the spine to provide essential structural stability and promote osteogenesis. The fibula and the rib are the usual donor bone sites. At least, a partial vertebrectomy is often performed, particularly in the thoraco-lumbar spine. The fibular or rib graft is inserted into surgically prepared notches at the superior and inferior vertebral levels. Anterior vertebral cortex is preserved to prevent graft displacement.

f) Clothespin (H) graft: This type of bone grafts has been used particularly in cervical spine fixation. Iliac crest full-thickness bone graft strips are slotted both proximally and distally and then placed between spinous processes. Additional cancellous bone is usually inserted as well. The laminae are decorticated to provide a large osseous surface for osteogenesis.

g) Vascularized graft: Free vascularized bone grafts are in extension of microvascular transplantations technique. The indications for vascularized graft are large segmental bone defects from trauma or aggressive tumour resection, mandibular and calvarial reconstruction, composite defects (bone, skin and muscle) and refractory non unions. For this technique, vascular anastomosis is performed after fixation of the graft in the recipient bone. These vascularized autograft are associated with improved survival of osteogenic cells and more active participation in osseous incorporation.

h) Orthotropic graft: The site of graft placement in appropriate site

i) Heterotopic graft: The site of graft placement in anatomically inappropriate site.

j) Fresh graft: Fresh grafts are transferred directly from the donor to the recipient and are almost always autograft.

k) Preserved grafts: bone grafts may be preserved by freezing, freeze-drying, irridation, autoclaving or chemical preservation. Cell death results from all these techniques and the graft functions mainly as space filler and a scaffold for in-growth of new host bone. Maintenance of sterility is of primary concern with any of the preservation methods.

Allograft (homograft)

Autogenous bone graft is currently the most effective material available for most situations necessitating osseous implants. This is because of the superior capacity to promote osteogenesis, the optimal ability to become incorporated and the lack of immunologic reaction characteristic of allograft. However, autogenous bone grafts also display major disadvantages in several clinical situations. These limitations include (1) insufficient amount of graft material available for surgical requirements when dealing with large osseous defects (2) significant post operative risk of morbidity at the donor site (structural weakening, infection, hemorrhage, pain, nerve injury and persistent deformity) and (3) the inability to mold the autograft in a shape to provide optimal function.

The allograft (homograft) is a tissue that is transferred from one individual to another individual of the same species. The limitations associated with the procurement of autograft for bone grafting can be overcome by the use of allograft. However, disadvantages associated with preserved allograft are (1) it is foreign and therefore may excite immune response in patients, (2) it is dead and cannot itself contribute to osteogenesis (3) it may not be as osteoinductive as fresh autograft and (4) it may not remodel rapidly. Therefore, presently the research is directed towards treating the bone grafts by different methods so as to preserve their osteoinductive and osteogenic property and to reduce the antigenicity.

The bone can be processed by various methods for preservation before storage which are boiling, autoclaving, deproteinizing and merthiolating methods to reduce the immunogenecity of alloimplant, but they also retard the capacity of osteoinduction. Treated bone is resistant to revascularization and remodeling by the host. Because of these experimental observations as well as extremely poor clinical results, these methods of preservation are rarely used. Bone decalcified by hydrochloric acid retains its osteoinductive capacity. The decalcified bone grafts have little strengths and cannot be used as a weight bearing strut. The decalcified allogenic bone grafts are preferred over boiled, autoclaved, deprotenized and merthiolated grafts. The bone matrix of the grafted tissue also influences the reconstructive properties of the host bone by the process of assimilative induction. It is further reported that highly cellular grafts (cancellous) with low matrix proportion get less opportunity for healing as compared to low cellular grafts along with high matrix proportion cortical.

Decalcified allogenic bone grafts

The histiocytes, foreign body giant cells and inflammatory connective tissue cells are stimulated by the degradative products of dead bone matrix to grow in and repopulate the area of an implant of decalcified bone. Histiocytes being numerous may transfer collagenolytic activity to the substrate; to cause dissolution of the matrix. This process is followed immediately by new bone formation by auto-induction.

The organic matrix of the bone and dentin, even in the absence of living cells, are found to possess a bone induction principle which is termed as bone morphogenetic protein (BMP). Its being chemotactic protein had a significant role in the migration of osteoprogenitor cells to the site of bone repair. The BMP extracted from the demineralized bone matrix is heat labile, hydrophobic non-species specific glycoproteins, which acted as a potent chemotactic for osteoblasts like cells but not for monocyte. The early response involves the appearance and proliferation of mesenchymal cells at the site of implantation followed by cartilage and bone formation. Acid decalcified bone induces the osteogenesis in a muscle and fascia, but induction is inhibited in spleen, liver and kidney. Studies on decalcified bone induced osteogenesis suggested that for bone induction to occur in soft tissues, three conditions must be fulfilled which are an inducing agent, the osteogenic precursor cells and an environment permissive to osteogenesis. The presence of osteogenic inhibitor in spleen, kidneys and liver might be responsible for not allowing osteogenesis in these organs. Decalcified bone matrix proves to be highly osteoinductive and readily osteoconductive material. Decalcified grafts do not evoke any appreciable local foreign body immunogenic reaction.

Attempts have been made to sterilize decalcified bone grafts by using various methods. Cobalt-60 used to sterilize demineralized bone, but the deleterious effects of radiation, even at much lower dose caused the destruction of BMP of bone. Sterilization is also attempted by aseptic collection and freeze drying. But this too have limitations which are sampling problem for monitoring sterility, degradation of BMP by neutral proteases and transplantation antigen during freeze- drying. The delayed hypersensitivity reaction induced by aseptically collected freeze dried bone and destruction of BMP caused by radiation sterilization is avoidable by sequential chemo-digestion and chemo-sterilization of bone, that preserve the maximum morphogenetic potential while transferring a minimum quantity of alloantigen. Avoidance of harvesting procedure and the potentially unlimited supply of banked material are their major advantages. It also provides important tool for bridging of large bone defects and repair of high risk fractures. Histopathological studies reveal that there are three phases in the allograft incorporation: degenerative phase, reparative phase and differentiation phase. When these phases are compared to autograft incorporation, it is found that (1) the initial phase of allograft incorporation is longer than the initial phase of autograft incorporation (2) the cells from granulation tissue differentiate into fibrocytes and giant cells in differentiate phase of allograft and subsequently into osteoprogenitor cells. However, no change is noticed in reparative phase. The slight to moderate amount of necrotic tissue is seen at that grafted site during allograft incorporation.

On the basis of radiological observations the fate of implanted freeze-dried-cortical allograft are divided into 3 phases. During the first 3-6 weeks, the graft appears dense and do not change morphologically. In the second phase, which lasted for a few days, the occurrence of peripheral graft resorption and replacement with host bone is seen, and in third phase, which lasted for several years, there is the continuation of bone resorption and replacement process. This process of gradual allograft replacement has been referred to as creeping substitution. The decalcified bone grafts is found to be superior to autologous bone grafts with regards to their osteoinductive potential on the basis of radiography, histopathology and radioisotope uptake studies.

Frozen allograft

The bone grafts can be stored by deep freezing without elaborate equipment and minimal personal involvement. The frozen allograft in large animals shows no reactions and vascular penetration into the graft is superior as compared to fresh allogenic, autoclaved and merthiolated allogenic bone grafts. Frozen allografts are less immunogenic than fresh allografts and retain the ability to induce osteogenesis. Although revascularization and remodeling are delayed compared with fresh autograft, resorption and osteoconduction occur more rapidly and completely in frozen implants than in fresh allografts. The recommended temperature for storage of frozen graft is -70^0 C. Autolysis may occur at warmer temperature (-10°C) since bone does not freeze at these temperatures. However, successful use of alloimplant stored in conventional freezers (-10°C) has been reported. Deep frozen bone grafts are rapidly resorbed and replaced by the new bone followed by autoclaved, decalcified and inorganic bone grafts in decreasing order. The bone grafts could be stored at -196°C without significant changes in the biological and physical properties.

Composite bone grafts

i. Recombined grafts of bone and marrow

The beneficial effect upon osteogenesis of impregnating bone grafts (homograft) with autologous red marrow is well documented. Autologous marrow transplanted to heterotopic sites has been known to form new bone. The presence of osteogenic precursor cells in marrow has been extensively evaluated. Bone marrow is composed of myelogenous and stromal cells, which although integrated in the marrow tissue, are morphologically and functionally different and extremely difficult to separate from each other. In bone marrow transplants, haematopoietic cells die shortly after transplantation, and there is an overgrowth of fibroblasts. These connective tissue elements are reversible modulations of one and the same cell type, a stem cell. They can differentiate and still have an ability to proliferate, redifferentiate, and give rise to bone formation. Demineralized bone acts only as framework for mineralization and the transplanted marrow cells are essential for osteogenesis.

ii. Plasma coated demineralized bone graft

The active components of demineralized bone matrix are the low molecular weight proteins, or bone morphogenetic protein (BMP), which is known as bone induction principle. In addition to protein of bone matrix, fibronectin, a cell surface protein is also present in plasma. Plasma fibronectin has been demonstrated on the surface of particles of demineralized bone matrix. The localization of fibronectin on the surface of the particles has been shown to be necessary for ossification induced by demineralized bone matrix and that antibodies to fibronectin prevent the normal process of endochondral ossification around the demineralized bone matrix. Fibronectin is a large, adhesive glycoprotein with many biological properties. Both the cellular and the plasma forms of fibronectin can bind cells and mediate their adhesion to collagen. The graft of demineralized bone matrix that have been treated with autologous plasma is called plasma coated demineralized bone matrix or augmented plasma coated demineralized bone matrix. Experimental evaluation of the healing grafts of demineralized bone matrix demonstrate that pretreatment of the demineralized bone matrix with either citrated plasma or fibronectin accelerate the appearance of the formation of bone on radiographs and enhanced the continuity of the bone that is formed.

2. Bone Substitutes

Ceramic implants

Ceramics used in orthopedic surgery are classified as bioactive or inert according to the tissue response when implanted in an osseous environment. The bioactivity of a material can be defined as its ability to bond biologically to bone. An inert ceramic merely elicits a minor fibrous reaction. In clinical practice, inert fully dense ceramics are used as bearing in total joint replacements because of their exceptional resistance to wear properties. By contrast bioactive ceramics are employed as coatings to enhance the fixation of a device or a bone graft substitutes because of their osteoinductive properties.

The size and regularity of the pores is a critical factor in the design of ceramic implants for potential clinical use, not only for tissue penetration and bone ingrowths, but also for physical strength. In order to allow tissue penetration and bone formation within the ceramics, large pores are essential. It is found that pore size greater than 100μm is necessary to allow the ingrowths of newly formed osseous tissue.

The most obvious drawback in the orthopaedic use of ceramics as scaffolding and fixation material is their inherent brittleness. Although recent advances in technology and engineering design have decreased the problems associated with brittle fracture, it still remains as the single most detrimental characteristic of ceramics for structural use. Therefore the availability of a ceramic implant material, which would be compatible with the body tissue and fluids while slowly dissolving as it is infiltrated by new bone, is of significant importance. Subsequent replacement of the ceramic with new bone would alleviate the ceramic strength problems, and the anticipated increase in mineralization rate would minimize the time required for the patient to remain inactive.

Bioactive ceramics

Bioactive ceramics are osteoconductive, act as a scaffold to enhance bone formation on their surface and are used either as a coating on various substances or to fill bone defects. An osteoconductive material can only elicit bone formation in an osseous environment, whereas, an osteoinductive substance (eg. Bone graft) can promote bone formation even in an extra-osseous situation. Different bioactive ceramics are tricalcium phosphate (TCP), aluminum-calcium-phosphorus oxide (ALCAP), calcium hydroxyapatite (HA) and Zinc-calcium-phosphorus-oxide (ZCAP) etc.

Bioactive bone cement

During the last 10 years bone cement materials have grown in popularity and are very promising osteoconductive substitutes for bone graft. These materials have been introduced in surgery to replace polymethyl methacrylate (PMMA) in order to avoid complications related to PMMA debris and to enhance fixation of the joint prosthesis. They include calcium phosphate based bone cement and glass ceramic bone cement. A strong cement-bone interface is obtained by formation of HA at the surface of the cement; moreover, calcium phosphate cements are resorbable and are progressively replaced by newly formed bone. They are prepared like acrylic cements and contain a range of powders such as monocalcium phosphate, tricalcium phosphate and calcium carbonate, which is mixed in a solution of sodium phosphate. These cements are produced without polymerization and the reaction is nearly non-exothermic. The final compounds have strength of 10-100 Mpa in compression while 1-10 Mpa in tension, although very weak under shear forces. These composites are currently used in orthopaedic in the management of fractures.

Zirconia (Y-YZP) ceramic

This ceramic implant is introduced in surgery in the manufacture of femoral heads for total hip replacement because of its higher strength and toughness, which would reduce the risk of fracture. Pure Zirconia is an unstable material showing three different crystalline phases: monoclinic, tetragonal and cubic. Zirconia metal head should articulate only against polyethylene sockets since both Zirconia against alumina and Zirconia against Zirconia have been shown to produce catastrophic rates of wear *in vitro*. Zirconia-on-polyethylene has demonstrated similar rates as alumina-on-polyethylene *in vitro*. Zirconia ceramics have several advantages over other ceramic materials. Due to transformation toughening mechanisms operating in their microstructure that can give to components made out of them. Zirconia is much stronger than alumina and thus risk of catastrophic failure from stresses is lower with Zirconia. Alumina and Zirconia are the bio-inert ceramics used for prosthetic devices.

Aluminium –calcium-phosphorus oxide (ALCAP) ceramic

The components of ALCAP ceramics are aluminum oxide, calcium oxide and phosphorus pentoxide in a ratio of 50:34:16 by weight. The density of ceramic implants plays a major role in their degradation after implantation. The ALCAP ceramic implants of various densities can be fabricated by varying the particle size, time and temperature of sintering. The degradation of ceramic implants is also influenced by the ratio of aluminum, calcium and phosphorus oxide and in the starting particle size besides time and temperature of sintering.

ALCAP ceramics are unique in the sense that they provide a multipurpose crystallographic system, where one phase can be more rapidly resorbed than the others. The porous nature of the ceramics allows induction of new bone growth into the implanted ceramic. Optimum pore size ranged from 75-150μm in diameter seemed necessary for the ingrowths of bone vascularisation and adequate oxygen supply.

ALCAP ceramics are tested *in vitro* by physiochemical studies in human plasma, hemolysis tests in human blood and tissue culture tests consisting of the Ames test and agar overlay. *In vitro* tests are intramuscular test and histopathological tests that included hematology, plasma chemistry, urine analysis and measurement of aluminum, calcium and phosphorous in muscle, bone, plasma and urine. Resorption of calcium and deposition of inorganic phosphates in the implanted calcium-aluminate ceramics suggest that ions are being exchanged with body fluids. The aluminum, calcium and phosphorus resorbed from ALCAP ceramics does not accumulate in toxic amounts in brain, bone, heart, liver and spleen, and the main route of removal from implant is via the surrounding tissue, plasma, and kidney and through excretion in the urine.

Histopathological, radiological, gross and triple fluorochrome labeling studies (Maiti and Singh, 1995, 1996) reveal that decalcified composite allografts have better osteoinductive and osteoconductive properties than ALCAP ceramic implant. Composite grafts of ALCAP and autogenous bone marrow shows slow healing as evidenced by persistence of implant pieces and absence of complete osseous union at the implant site even after 90 days of implantation, as compare to composite grafts of frozen decalcified allografts with fresh autogenous bone marrow. Resorption of ceramic implants is very slow as compare to bone grafts.

Tricalcium phosphate (TCP) ceramic

Tricalcium phosphate has a nominal composition of $Ca_3 (PO_4)_2$. The common mineral name for this material is whitelockite. It exists in two –crystallographic forms, alpha and beta whitelockite. In general, it is found in the beta form. The β- form is most stable with calcium-to –phosphate atomic ratio of 1.5. These ceramics are considered to be resorbable since the rate of biodegradation is higher when compared with HA. Degradability occurs by combined dissolution and osteoclastic resorption. The principal clinical application of TCP has been in dentistry. Powdered TCP has been used for initiating apical closure in teeth and for treating periapical defects and to augment alveolar ridge defects when aluminum oxide tooth roots are implanted.

Hydroxyapatite ceramics possess a higher osteogenic potential than do the whitelockite materials. In general, rate of absorption of tricalcium phosphate is determined by microporosity of the specimen, the higher its rate of degradation, however, ingrowths of new bone into tricalcium phosphate is not always determined by the micro-porosity of the ceramic. Perhaps the most consistent and desirable property of tricalcium phosphate, as well as other calcium phosphate ceramics is biocompatibility. No foreign body reaction, local inflammatory response, or systemic toxicity has been reported. The pathways of calcium and phosphate ions have been traced in urine without any abnormal results. The most remarkable property of calcium phosphate ceramics is their ability to bond directly to bone, although the existence of this bond is yet not fully understood. The testing of its biocompatibility shows that it is devoid of systemic or local toxicity and free of irritation or foreign body response in tissues and did not result in hemolysis or mutation. The new bone is readily grown into its pores with direct contact with the implanted TCP material. The TCP-plasma sprayed titanium rods exhibit significantly higher pull-out strengths than unsprayed contralateral rods, suggesting the bonding of bone to the TCP layer. It also shows significantly greater osseous tissue response in the TCP-sprayed rods than unsprayed contralateral controls. The material may eventually be approved for use in orthopedics.

A significant limiting factor in the application of TCP is its mechanical properties. Like most ceramics, it is brittle and has low impact resistance. Porous TCP has a compressive and tensile strength similar to, but on the average lesser than cancellous bone, depending on the manufacturing process. Porous TCP implants have greater solubility than hydroxyapatite and are resorbed more rapidly. Composites of TCP with a porous hydroxyapatite (HA-TCP) structure encourage bone ingrowths and new bone formation. Commercial products of β-TCP are Synthograf, Augmen and Himed-TCP.

Calcium hydroxyapatite (HA) ceramic

Hydroxyapatite is the major mineral of bone. The nominal composition of this mixture is $Ca_{10}(PO_4)_6(OH)_2$. Synthetic hydroxyapatite is chemically and crystallographically identical to the natural occurring hydroxyapatite of bone and it is an extremely compatible material with calcium –to-phosphate atomic ratio of 1.67. It is introduced as a bone graft substitutes because its formula is similar to that of inorganic mineral phase of bone. Biological HA, however, is Ca-deficient and a carbonated apatite. The bonding mechanism of HA to bone, although not completely understood, seems to be due to attachment at the surface of the HA of osteogenically-competent cells, which differentiate into osteoblasts. A cellular bone matrix is then formed at the surface of the HA. An amorphous area is present between the surface and the bone tissue containing thin apatite crystals. As maturation occurs, this bonding zone shrinks and HA becomes attached to bone through a thin epiaxial layer, resulting in a strong interface with no layer of fibrous tissue interposed between the bone and HA. Bone formation grows from the surface of the HA towards the centre of the pores.

Calcium phosphate ceramics (hydroxyapatite and tricalcium phosphate) in a particulate form are placed in gaps between the implant and the host bone. Hydroxyapatite is neither "osteogenic" nor "osteoinductive". It is however, "osteophilic" and "osteoconductive". These materials appeared to be osteoconductive, i.e. they served as scaffolding for bony in-growth. In addition, they act as extenders of autogenous bone graft. The main disadvantages, which have limited the clinical application of HA as a bone substitute, are related to the brittle nature and poor tensile strength of the material. The synthetic hydroxyapatite acts as a trellis for the ingrowths of vessels and the subsequent deposition of new bone.

Several forms of hydroxyapatite have been made for clinical uses; solid ceramic blocks, porous blocks and particulates. Porous blocks of HA have been used for craniofacial reconstruction. Solid, dense forma have been used to repair bone defects in animals and as tooth root implants following extraction. Particulate HA is used in oral surgical procedure to augment the alveolar ridge and in periodontal repair. Hydroxyapatite (HA) coating is now widely used on femoral prosthesis and on sockets as a means of fixation in order to avoid complications related to hip joint transplantation. It is usually applied by plasma spray. Addition of silica in hydroxyapatite ceramics (HA-Si) causes early degradation in the body fluid. The main disadvantages, which have limited the clinical application of HA as a bone substitute, are related to the brittle nature and poor tensile strength of the material. Commercial products of calcium hydroxyapatite are Calcitite, Alveograf, Osteograf, Bioapatite, Bioroc and Mitsubishi-HA.

Zinc-calcium-phosphorus oxide (Z-CAP) ceramic

ZCAP organic acid composites fabricated by ZCAP-malic acid +15%$CaOH_2$ and ZCAP-cysteine. Postsurgery examination shows that both composites are biocompatible and non-toxic, and allowed ingrowths of new bone similar to TCP and HA. Injection of ZCAP-cysteine composite into the experimentally induced vertebral fractures in pigs has no significant effect on the animals. Radiographic evaluation shows that resorption of the composite and healing of the fracture successfully.

Calcium aluminate/Alumina ceramic

The prototype of the oxide ceramics, aluminium oxide, alumina, Al_2O_3 or corundum ceramic is the most widely used ceramic material for permanent surgical implant. Corundum is the colorless version for sapphire and ruby and has a melting temperature of 2050^{0}C. The use of this ceramics is increasing rapidly due to the extremely low wear rates in their articulation against the same ceramic and against polyethylene sockets where the wear rates is reduced by a factor of about 2 as compared to Co-Cr-Mo alloys. Sintering alumina powder at temperature between 1600^{0}C and 1800^{0}C obtain dense alumina of surgical grade.

The resultant material is in its highest state of oxidation, allowing thermodynamic stability, chemical inertness and excellent resistance to corrosion. There are two types of calcium aluminate: monophasic and polyphasic. The polyphasic calcium aluminate ceramics, which are resorbable, hydrolyze in a living system and undergo an increase in pore size, allow more bone infiltration and new bone formation. It is suggested that the addition of phosphate to calcium aluminate ceramics would increase their resorbability without sacrificing their strength. Alumina is used for joint replacement, prosthetic devices and filling bone defects. The ability to control porosity and solubility of alumina ceramics offers the possibility of use as drug delivery systems.

Coralline hydroxyapatite

Coralline hydroxyapatite is a synthetic material manufactured from marine coral and it has a natural trabecular structure similar to cancellous bone and provides osteoconduction. The implant material used in orthopaedic is derived from the mineral component of marine coral and is called coralline. It is mechanically weak and slow in absorption, but it is biocompatible and safe to use as small bone defect filler. It supports growth of both osteoblasts and bone marrow elements. The porosity and interconnectivity of the pores of coralline hydroxyapatite are important for bone ingrowths. These ceramics are used with autograft to act as bone graft extenders. Several bone growth factors are added to coralline ceramics, which stimulate the rate of bone formation.

Bioactive glass ceramic

Bioactive glasses are first developed by Hench and Wilson in 1984 and have a vitreous structure. They bond chemically to bone. Since the discovery of stable SiO_2-Na_2-O-CaO ceramic formulations, most silicate ceramics used have a SiO_2 content of 65 wt. % or more. These ceramics are more commonly called glass. These 65wt. % silica glasses are extremely bioinert and contained few other desirable qualities. Unfortunately, this type of glass is still very weak and brittle. The addition of P_2O_5 to the SiO_2-Na_2-O-CaO matrix makes the glass extremely bioactive. The first bioactive glass is Bioglass 45S5. Bioactive glasses are available in different trade name, such as "Ceravital", "Bioglass 45S5"; they are so bioactive that within one hour of implantation an HCA layer (precursor to bone ingrowths) nearly 500 nm deep is formed. It has also been reported that soft tissue growth on the surface of "Bioglass 45S5" has been observed. Bioactive glass ceramics are osteoconductive and biocompatible and can be used as a bone substitute material.

3. Bone bioimplants/biomaterials

The term bone bioimplants or orthopaedic biomaterials are synonymous with "materials used for internal fracture fixation'. The ultimate aim of bioimplants as applied to fracture fixation is to restore the structural integrity of the injured bone. This will depend upon a complex interplay of materials properties, device design and physiological requirements. The important considerations are site and type of fracture, operative approaches, rapidity of bone healing and the postoperative care.

The requirements of materials for internal fixation fall into three categories: clinical, manufacturing and economic. The clinical requirements are that the materials have suitable mechanical properties to fulfill its function of fixation and maintenance of fracture reduction. Additionally, it should not degrade in response to the corrosive conditions of the biological environment leading to releasing potentially harmful degradation byproducts locally and systemically.

Manufacturing requirements include that the materials have properties which permit fabrication in the optimum design configuration. The cost requirements, although not as critical in human orthopaedic must be considered in veterinary orthopaedic.

The mechanical properties of orthopaedic biomaterials are described in terms of their deformation or strain produced by an applied stress. The required mechanical properties of implant materials are determined by static or low cycle short term *in vitro* materials-testing methodology. The mechanical properties of "fatigue" "creep" and "stress relaxation' reflect time dependent materials behaviour and are important mechanical parameters in the selection process.

The most important non mechanical requirement of orthopedic biomaterials is "inertness". Ideally an implant material should not degrade. However, such a state is not achievable and therefore, a relative degree of implant degradation should be considered acceptable. The degradation process, however, should not impair significantly the mechanical strength of the device, nor allow the release locally or systemically of byproducts. Other requirements of metallic and non metallic orthopaedic biomaterials include manufacturing, fabricability and cost.

Stainless steels alloys: This alloy primarily made of iron, chromium and nickel. It is protected from corrosion primarily by the addition of chromium. The chromium forms a passivating layer of chromium oxide which aids in preventing oxidation of the iron (corrosion). Molybdenum is also added to the alloy to improve corrosion resistance in saline and acidic environment particularly to pitting and inter—granular attack. Addition of molybdenum is also hardness the alloy. The 316 L stainless steels alloy used for surgical implants contains 17-19% chromium, 12-14% nickel and 2-3% molybdenum. The carbon content in 316L stainless steels should not exceed 0.03% to maintain good corrosion resistance. A number of high strength stainless steels have been developed. Among them 22Cr-13Ni-5Mn (22-13-5) stainless steels have ideal properties for implant application. Stainless steels are popular for use as fracture implants, because the base materials are relatively inexpensive. The 316 L is used in hip stems and fracture fixation devices such as intramedullary pins, spinal pins, hip screws, bone plates and bone screws. The 22-13-5 is used in hip stems and fracture fixation devices such as hip screws, bone plates, spinal rods, and intramedullary nails.

Titanium alloys: Titanium alloys are extremely biocompatible and inert, with a modulus of elasticity closer to that of human bone than many other orthopaedic implant alloys. Due to their strength, flexibility and biocompatibility, titanium alloys are often used for joint and bone implants. Among titanium alloys, Ti-6AI-4V (containing 6% Aluminium and 4% Vanadium) has been widely used in the manufacture of orthopaedic implants. It is used in hip stems, knees, shoulder joint, elbow joint and fracture fixation devices such as intramedullary nail, hip screws, plates and bone screws. Due to their high cost, its uses in veterinary surgery is limited, however, they are the material of choice for human patients having known hypersensitivity reactions to any of the constituents of stainless steel or cobalt chromium alloys.

Cobalt alloys: The Co-Cr-Mo alloys are in use in the manufacture of orthopaedic implants. The cobalt based alloys generally contain about 30 wt% of Cr and about 7 wt% of Mo. All Co-Cr alloy casting used in orthopaedic implants are fabricated by investment casting in which a wax model is made and a ceramic shell is build around the wax model. When wax is melted away a ceramic mold has the shape of implant. Casting produced by this method exhibit adequate strength, good corrosion properties, and excellent wear resistance for implant application. The alloy is used for implantable devices having a fixed configuration, such as total hip prosthesis. Co based alloys used in hip stems, shoulder joints, elbow joints, knee joints, and fracture fixation device such as bone plates, hip screws and bone screws.

Titanium-Zirconium alloy: It's a new titanium alloy containing 13% niobium and 13% zirconium. These three elements make this alloy the most biocompatible alloy developed for orthopaedic application. The elastic modulus of this material is lower than any other metal alloy for implant application. Clinical utilization of the material includes bone plates and bone screws.

UHMWPE: The ultra-High-Molecular Weight Polyethylene has been introduced in hip cup by Sir John Charnley and since then it has been the bearing material of choice in orthopaedic joint prosthesis such as artificial hip, knees and shoulder. They are used as articulating surfaces against ceramic components in joint replacements.

PMMA: Poly methyl methacrylate is the only material which is likely to be handed and prepared by the surgeon in a professional rather than in a do it yourself role. The main commercially available methyl methacrylate is CHW bone cement, Simplex-B, Palacos, Technovit 609I. The basis of acrylic cement is methyl methacrylate which can exist either in a liquid (monomeric) form or solid (polymerized) form. In routine method the solid and the liquid components are mixed together in sterile condition until the semi fluid material reaches a doughy stage. Mixing of liquid and powder together brings about the polymerization of liquid monomer, which then binds together, the previously polymerized powder. The doughy material is then rolled in the palm of the gloved hand for ½ to 1 minute. Total time takes around 2-3 minutes. The completion of polymerization is characterized by non-adherence to the gloved hand and rise of temperature of the polymerized mixture. PMMA polymers are extensively used as biomaterials in dentistry and orthopaedic. Cementing with polymethyl methacrylate is a common means of fixing total hip prosthesis. PMMA has been used to design external fixation splint in transfixation pinning. It has been used as external splint after resection of tumour and auto grafting in mandible in human patients. It has been tested as acrylic side bar for external skeletal fixation in ruminants. Methyl Methacrylate has been used as an adjunct for internal fixation in unstable comminuted fractures and pathological fractures in osteoporotic patients. It has been used in conjunction with K-nail to fix its distal end to the femur in aged patients. Since PMMA is brittle, many methods have been employed to reinforce it with metal wires and plates to increase its transverse strength. The addition of saline treated glass fibers increased its transverse strength. Acrylic bone cements have been used widely in patients with 10%failure. Therefore, better alternatives to acrylic cements are yet to be established. However, the optimization of the preparation and the implantation methods seems to yield an improved *in vivo* behaviour. A good knowledge of the deformation and failure mechanisms would make it possible to design the optimum microstructure, which would fit the requirements of the mechanical loading conditions.

Summary

Bone grafting procedures constitute a vital surgical approach to the management of a variety of orthopaedic abnormalities. Besides providing mechanical support at the reconstruction site, an ideal bone graft must possess osteogenic, osteoinductive and osteoconductive properties. The availability and the ease of application are other criteria which can support the candidature of a graft to be an ideal bone graft. None of the available type of bone grafts possesses all these properties, therefore, a single type of graft can not be recommended for all types of orthopedic abnormalities. The indications of different grafts vary from condition to condition. A search for an ideal bone graft is on and may continue till time to come.

The excellent biocompatibility and outstanding biological properties of bioactive as well as inert ceramics have encouraged their use in total joint transplantation and bone reconstructive surgery. Progressive improvement in the manufacturing process has allowed the production of reliable ceramic materials. At present, there are obviously many factors, such as brittleness and low mechanical strength of ceramics that are to be considered in determining whether it is feasible to replace bone by ceramics. An ideal porous biodegradable ceramic to fill a bone defect is yet to be developed. In the near future, ceramic substitutes for bone grafts will probably be used in association with osteoinductive materials like bone morphogenetic proteins or mesenchymal stem cells to further accelerate bone formation.

Bone bioimplants used for orthopaedic devices should possess good biocompatibility, adequate mechanical properties, and sufficient corrosion resistance and should be manufactured at a reasonable cost. Because of their unique properties, metals have found successful application in the field of veterinary orthopaedic as prosthetic and fracture fixation devices. Metallic bioimplants are normally used for load bearing members as pins, plates and femoral stems etc. The metallic materials must have sufficient fatigue strength to endure the rigors of daily activity. The three major contemporary groups of metals are used in the majority of veterinary orthopaedic: stainless steel, titanium and its alloys and cobalt based alloys.

Bone bioceramic: 1-ALCAP ceramic block; 2-Tri-Calcium-Phosphate (TCP) ceramics; 3-Hydroxyapatite (HA) ceramic powder; 4. Hydroxyapatite-Silica (Ha-Si) ceramic block; 5-Hydroxyapatite-TCP ceramic block

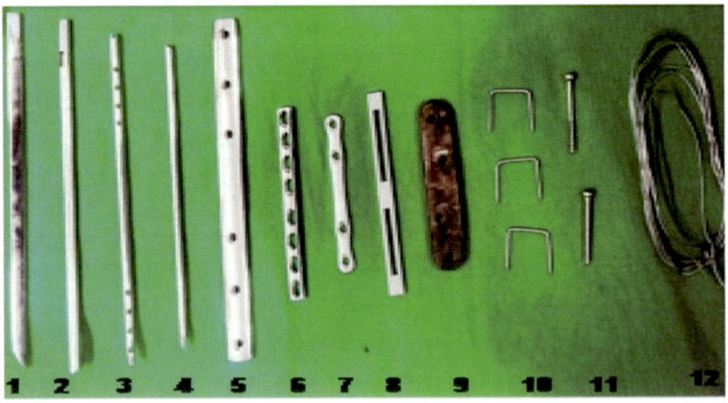

Bone bioimplants: 1-Kuntscher nail (V-nail); 2- Kuntscher nail (K-nail); 3-Interlocking nail; 4-Intramedullary Pin; 5- Heavy Duty bone plate; 6- Dynamic Compression plate (DCP); 7-Shermann bone plate; 8- Egger's bone plate; 9- Horn plate; 10- Bone staples; 11-Bone screw (above-cancellous, below-cortical); 12- Orthopaedic wire.

5

Internal Fracture Fixation in Small Animal

Repairing fractures in small animal's presence challenges to the veterinarians in their day to day service. These challenges include properly repairing fractures, each of which is unique, choosing the right fixation device, keeping an extensive inventory of costly instruments, and providing high standards of care for fracture repair. But fracture repair also gives veterinarians the opportunities to do something a little unusual in daily practice and to experience the immediate gratification of easing pain and promoting activity in injured animals.

Successful management of fractures mainly depends upon early diagnosis, type of fracture, timing of surgery and selection of proper method of fixation. The fixation of fractures can be achieved either by external or internal fixation methods. An ideal method of fracture fixation is one that can maintain early anatomical reduction, minimal soft tissue trauma, rigid stability and early motion of bone and joint. Restoring a fractured bone's normal shape is important for return to function because it contributes to normal alignment of adjacent joint surfaces. Whereas, rigid stability of fracture fixation eliminates fragment movement and reduces pain. Soft tissue trauma adjacent to fractured bone should be minimum, which play a vital role in early vascularisation and prevents infection. Finally, early return to function maintains muscular tone of the limb allows normal weight mobility and joint nutrition and uses normal weight bearing forces to maintain bone density. These can be achieved with use of internal fixation devices.

1. Fracture diagnosis

Following methods can help in diagnosis of fractures in terms of type of fracture, site, and bone involved, time lapses and extent of soft tissue involvement.

- History- How and when the fracture occurred?
- Physical examination- Assess the animal for abnormalities other than fracture, like occult injuries to the thorax and abdomen (vehicle injury).

- Signs- Fractured animal often present with limb dysfunction, pain, instability, overlying soft tissue trauma, and abnormal posture or limb position and crepitus.
- X-ray- A radiographic examination (at least two divergent views, particularly at 90 degrees to one another) will confirm the type, location, and configuration of fracture.
- CT/MRI- If available, computed tomography or magnetic resonance imaging is useful for detailing fractures.
- Bone scan- Bone scintiagraphy is useful in diagnosing non-displaced fractures and fractures of distal extremity sesamoids.

2. Fracture biomechanics

For proper repair of fractures, a complete understanding of the forces acting on a bone during fracture healing is important. Force is defined as a vector quantity having magnitude and direction. The forces and movements (rotational forces) applied to bone result in loading modes that require understanding to appreciate the biomechanics of fracture etiology and fracture repair. Each fracture repair must allow for and counteract all forces to allow optimal healing. The five basic forces on along bone are tension, compression, bending, shear and torsion (rotation). Tensile forces act along the long axis of the bone and attempt to lengthen the bone. These forces cause transverse or avulsion fractures. Compressive forces also act along the long axis of the bone and attempt to shorten the bone, causing impaction or depression fractures. Bending forces applied to a specific focal point on the bone, cause transverse or short oblique fracture. Shear force is transmitted parallel to but not along the long axis of a bone. This force causes the fracture of bony promainces along the line of force or oblique fracture. Torsion is a twisting force applied to the long axis of the bone, causing spiral fracture.

3. Timing of surgery

It is important to consider the soft tissue response to injury when considering the optimal time to carry out internal fixation of a fracture. The local circulation will be disturbed causing hypoxia, acidosis and edema of the tissue, an ideal situation to become established. Surgery should either be undertaken before the circulation becomes compromised ideally when six hours (maximum 24 hours) or it should be delayed for between four to six days to allow the circulation to the soft tissues to become re-established. Urgent surgical treatment is needed for articular fractures, growth plate fractures and compound fractures. Early surgery also preferable for pelvic fractures and severely comminuted shaft fractures before muscle contracture makes reduction difficulty.

4. Factors for considerations in the choice of fracture treatment

- Location and configuration of fracture
- Forces acting on the fracture (biomechanics of fracture)
- Concomitant orthopedic and soft tissue injury
- Age, size and temperament of the patient
- Pre-existing disease
- Convalescent environment
- Owner compliance/expectation
- Availability of instruments/devices
- Expected performance level of the animal
- Economics

These factors must be considered during planning of a long bone fracture-repair.

5. Selection of appropriate method of fixation

The management of fractured differ according to the anatomical location of a long bone in small animal.

(a) ***Diaphyseal fracture***- Broadly divided into stable and unstable fractures. Stable fractures are the transverse, blunt oblique or greenstick fractures in which the fragments interlock and resist shortening. The fixation devices necessary is to prevent angular deformity and depending on site, this may done with an intramedullary pin, bone plate or Krischner splints. The unstable fractures are oblique, spiral and comminuted. The method of fixation should maintain length of bone and prevent angulations and rotation. The ideal way of doing this is to reconstruct the bone with multiple full cerclage wires in conjunction with intramedullary pin, interlocking pin and multiple interfragmentary screws in conjunction with a neutralization plate.

(b) ***Avulsion fracture***- In avulsion fractures, viz. fracture of the tibial tuberosity, the olecranon, greater trochanter, patella and os calcis, the fragment is distracted by the tensile force of the muscle tendon or ligament which inserts on the fragment. The fragment should be stabilized with Krischner wires or a lag crew used in combination with a tension band wire to counteract the tensile force acting on the fragment.

(c) ***Articular fracture***- In articular fractures, early open reduction is essential to allow accurate reconstruction of the joint surfaces. An arthotomy of the affected joint should be performed to permit accurate reduction of the articular surfaces and the removal of debris and hematoma from the joint.

Immobilization of the fragment is usually achieved with one or several transcondylar lag screws and a Krischner wire to prevent rotation.

(d) ***Metaphyseal fracture***- Metaphyseal fractures are most difficult to repair when close to a joint or severely comminuted. Stable internal fixation is preferable because it permits early ambulating and prevents decrease range of joint motion or in severe cases, fixation of the joint. Crossed Steinmann pins, Rush pins, bone plate and intramedullary pins can be used for internal fixation of metaphyseal fractures.

(e) ***Supracondylar fracture***- Fractures involving the region just above the condyles of the humerus or femur are frequently reported in immature dogs. Effective reduction of such fractures is difficult due to the small distal fragment and proximity of the fracture to the joint. While treating such cases, care must be exercised to maintain the joint function and allow early range of motion in addition to provide rigid fixation. Open reduction is essential to prevent caudal rotation of the condyles and malunion, various treatments like modified single or cross Steinmann intramedullary pinning (normograde) through the inter-condyloid fossa, crossed Krischner wires, Rush pins are practiced.

(f) ***Epiphyseal separations***- Early open reduction should be carried out and Krischner wires are used for fixation.

6. Antibiotic prophylaxis

Before fracture-fixation surgery, antibiotic prophylaxis should be provided. High blood concentrations should established by intravenous/intramuscular injection of antibiotic one hour before surgery. At least, a five day course of treatment should be given. Strepto-penicillin, Ampicillin, Enrofloxacin, Ceftriaxone and Tazobactam combination are used frequently. Poor asepsis, prolonged operating time, disruption of blood supply to the bone by rough handling of tissues and periosteum stripping or soft tissue attachment all predispose to infection. Antibiotic prophylaxis is no substitute for poor standards asepsis and surgical technique.

7. Principles of some common internal fixation methods

(A) Intramedullary fixation

Intramedullary fixation is a simple method of fracture repair, which is widely used in veterinary orthopedics. The intramedullary devices include the Steinmann pin, the Rush pin, the Krischner wire, the Kuntscher nail, and the interlocking nail. The intramedullary (IM) pins, with or without ancillary fixation devices are the most commonly used fixation device in general veterinary practices.

Ancillary devices

Full cerclage wires, hemi-cerclage wires, K-wires, lag screws, crossed K-wires.

(i) **Single IM pin fixation :** Intramedullary Steinmann pins are used in fracture repair to resist bending forces and maintain axial alignment of fracture. Primary indication of single pinning is in the treatment of transverse or blunt oblique fractures of the middle third of long bone. The pin can be placed in a normograde direction in any long bone and in a retrograde direction only in the femur, humerus or ulna. IM pins provide three pints of fixation in the medullary cavity: the point of insertion, the isthmus of the medullary canal and the distal cancellous bone. The bending strength of an IM pin is related to its diameter. A pin's diameter should be 70% to 90% of the medullary cavity's diameter in the diaphyseal region. As the pin lies within the medullary cavity it resists bending in all directions. During IM pinning, it should enter the cancellous at both ends of fractured bones and it should run the entire length of the bone. The use of threaded intramedullary pin should be avoided, as they tend to break at the thread/non-thread surface.

IM pin placement: Placement of IM pin in the medullary cavity can be achieved by either a normograde or retrograde fashion. With normograde placement, the pin inserted at either bone end and is directed towards the fracture sites. With retrograde placement, the pin is introduced at the at the fracture site and is directed towards the end of bone. After the pin is placed in the proximal or distal fragment, the bone is anatomically reduced and the pin is inserted into the opposite fragment. Normograde placement allows accurate placement of the pin entrance and avoids the most important regional structures whereas, retrograde placement has the advantages of simpler passage and easier selection of pin size in relation to the diaphyseal diameter. But it allows less accurate control of the exit point, possibly damaging joints, nerves or other soft tissues near the bone ends. Pin migration during healing suggests impending fixation failure. It is caused by movement at the fracture site or action of adjacent soft tissues on the pin end.

Limitation: The ability of IM pins to restrict torsion (rotational), tensile and compressive forces are limited by the lack of friction between the metal surfaces and the bone. IM pins should not be used as the primary fixation device to repair comminuted fractures unless all fragments can be reduced and stabilized with ancillary devices, otherwise an IM pin provides no longitudinal support or resistance to shortening forces and so collapse, rotation and non-union are to be resultant. IM pins are not good devices to use in the presence of sepsis. IM pin interfere with centrifugal blood flow through the reaming effect of their introduction and continued presence in the medullary cavity. IM pinning disturbs endosteal callus formation but it little interferes with the healing of the cortex and periosteum. Pin migration is the most common problem with IM fixation.

(ii) **Stacked IM pin fixation :** The use of two or more pining-stacked IM pining is indicated for transverse fractures when rotation cannot be controlled with a single IM pin or for fractures of large diameter bone when single pin of appropriate sizes are unavailable. For stacked pin fixation, insert each pin through a separate distinct proximal hole and into the medullary cavity, preferably allowing the opposite ends of the pins to diverse within the opposite metaphysis.

Limitation: Biomechanical and clinical evaluations indicate that the stack pinning technique still inadequately controls torsional (rotational) forces and often results in fracture non-union. As with single IM pinning, pin loosening or migration are other limitation of multiple pinning.

(iii) **Rush pin fixation :** Rush pins are solid, curved cylindrical pins with small diameters. Rush type fixation (cross-pinning) –Two pins are introduced from opposite sides of the epiphyseal cortex, cross near the fracture site, and then glance off the endosteal surface of diaphyseal bone. Rush type fixation are indicated for the reduction and fixation in young animals. Proper insertion of Rush pins forms a dynamic mode of fixation, with the spring like curve of the pins providing variable pressure on the endosteal surface as the animal bears weight. The presence of two opposing pins with separate insertion and diverging points in the diaphyseal provides a moderate degree of rotational stability. Rush pins have minimal effect on medullary blood supply due its smaller diameter.

Limitations: Rush pins are contraindicated in very young animals in whom the marrow cavity is large and the cortex is soft and may be easily penetrated by the Rush pin. When the point of pin pierces the opposite cortex, the dynamic mode of fixation is lost and the technique reverts to simple cross-pin fixation. The insertion or removal of Rush pins depends on a specially designed set of instruments. Other limitations include animal that has had a long episode of disuse atrophy in which the soft cortical bone may not be sufficient to support the Rush pin. Insertion technique requires practice.

(iv) **Interlocking nail fixation :** Interlocking intramedullary nails are relatively new fixation devices in small animal orthopedics. This nail is secured in position in the medullary cavity by proximal and distal transfixation screws which pass through IM nail and engage both the cortices. The primary advantage of inter-locking nails over the Steinmann IM pin is that interlocked screws add torsional and axial stability to the fractured bone, thus it provides the bending, axial and rotational stability in diaphyseal fractures and allows load sharing between the bone and implant. Fracture healing relatively easy to access.

Limitation: Non availability of right size and shape of nail, cost and necessary of intra-operative fluoroscopy for placement of the locking screws through the bone or nail or uses of Zig for this purposes, difficult to insert and remove are some limitations of this technique.

(v) **Kuntscher nailing** : V and "clover leaf" shaped of Kuntscher nailing is indicated for middle third, transverse or short oblique fractures of the humerus, tibia and femur when no longitudinal cracks or tissues of the bony cortex are present. It is designed to stabilize the bone fragments by filling the marrow cavity and contracting as much of the cortical bony surface as possible. In dog, it is used as three-point fixation device. In the initial point of fixation is the penetration of the nail through the proximal portion of the bone. The nail is then stabilized in the central portion of the fracture site, usually at the area of the isthmus of the marrow cavity and distally is anchored in the cancellous portion of the bone of distal diaphysis. It is best inserted from the proximal end of the bone distally. The nail must be measured to ensure the proper diameter and length. Following fracture reduction, the nail is inserted through the reamed trochantoic foramen (femur) and driven across the fracture site and seated in the distal fragment.

Limitation: It should never be used when there are longitudinal cortical cracks in either the proximal or distal fragments of the bone involved. For this technique, needs some specialized instruments for its placement.

Biodegradable IM rod: Rods formed from self-reinforced polygalactide or polylactide have been used for cancellous bone and in the diaphysis. In India, it is not available.

Primary fixation devices	Suggested uses
Intramedullary (IM single pin)	Transverse diaphyseal fracture of long bone
IM pin with full cerclage	Oblique or spiral fracture
IM pin with lag screw	Oblique or spiral fracture
IM pin with external fixator	Diaphyseal and metaphyseal fracture
Rush pins	Metaphyseal, physeal and epiphyseal fractures
Tension band fixation	Fracture of bone apophyses
Interlocking IM nail	Most diaphyseal fracture configurations
Bone plate	Most fracture configuration because of varied plate design

(B) Cerclage wire fixation

Cerclage wire is a metal wire loop that encircles the circumference of a bone. It is useful as ancillary fixation for fracture repairs in which IM pins, external fixator or bone plates are used. It provides rotational stability in oblique, spiral or comminuted fractures. It has been also used to hold bone grafts in place. 18 gauge (1.2cm thickness) wire is generally used in animals over 20 kg in weight whereas in animals under 20kg, at least 20 gauge (1mm thickness) should be used.

Principles of wiring

Full cerclage wire encircles bone completely. Hemi-cerclage wire pass through a tunnel in the bone at some point, which may provide more secure fixation but can be challenging to apply. At least two wires per fracture site should be used. The wire should be applied tightly. During multiple wiring, they should be spaced at least 1.0 cm apart. Cerclage wiring should be placed on perpendicular to the bone shaft and it should not be placed on the fracture line. In oblique fractures, the fracture lines should be at least twice the length of the diameter of the shaft to resist shearing force by the cerclage wire. Hemi-cerclage wire fixation is a technique in which wire is passed through holes drilled in adjacent fracture fragments. The wire is then tightened to form a "suture" like pattern. To prevent wire slipping (the most common complication of wiring) either notches the bone or use a hemi-cerclage wire in which the wire penetrates the cortex of the bone via drilled hole.

Limitation: The inherent danger of cerclage wiring in the management of comminuted fractures is that the fragments shift on load bearing, the wire may become loose, break or slippage into fracture line causing fracture instability. Fixation of bone fragment by cerclage also has deleterious effect on cortical blood circulation, subsequent bone healing, and callus bridging.

(C) Bone plates and screws fixation

Bone plates and bone screws are used to provide rigid stabilization in the surgical repair of simple and comminuted fractures. When properly applied, bone plates control all the forces acting on the fracture site.

Types of bone plate

- **Sherman plate**-weak plate and its use mainly limited to small dogs
- **Lane bone plate**- this is similar to the Sherman plate but far weaker
- **Venables bone plate**- there are no constrictions between the screw holes and the plate is strong and ideal for general orthopedics
- **Egger's Bone Plate**- the Eggers plate had two long slots that allowed the screw heads to slide and thus compensate for resorption of the fragment

ends. Widespread use of this plate is limited by its structural weakness and the resulting instability of the fixation.

- **Heavy duty Venables plate**- used in very large dogs
- **Burn bone plate**- this is a modification of both the Sherman and Venables type plates
- **Horn plate**- it is prepared from bovine horn. The plate of desired size and shape are prepared from the upper two-third of the horn (thick portion) by using bench grinder. Screw holes are then drilled at appropriate distance with the help of hand drill. It is completely inert, radiolucent, semi-rigid and light in weight. It can be sterilized in 70% alcohol and then autoclaving. This plate is used for different fracture fixation in the long bone of small animal. The main advantages over stainless-steel plate are easily available, economic, semi-rigid, and radiolucent, screw holes can be drilled whenever required and it can be contoured to the shape of fractured bone. The only limitation is less strong than stainless steel bone plate.
- **Veterinary Cuttable plate (VCP)**-the veterinary cuttable plate is also a unique and versatile type of plate, available in two sizes. The larger VCP is 1.5 mm thickness and 7 mm wide and it accepts 2 and 2.7 mm cortical bone screws. The smaller VCP is one mm thick and 7mm wide, and it accepts 1.5 and 2mm cortical bone screws. Both plates are 300 mm long with 50 holes and can be cut to the appropriate length at the surgery table (like horn plate). Another advantage of the VCP is that it can be stacked with other VCPs to achieve a variety of thickness and strengths.
- **Dynamic compression plate (DCP)** - it is the self compression plate due to the special geometry of the screw holes which allows the axial compression. The main feature of the DCP is the design of the screw hole (hemi-cylindrical slope) which is based on the spherical gliding principle. This enables the plate to be used as a self-compressing plate. Insertion of a screw in the load position will displace the bone beneath the plate towards the fracture site as the screw is tightened against the hemicylindrical slope of the screw hole as the screws are tightened the plate is placed under tension and the fracture site is compressed. If further compression is needed, another screw may be inserted on either side of the fracture. The remaining screws are placed using a neutral drill guide. The screw holes in a DCP are the weakest component of the plate but are somewhat protected when all screw holes are filled with screws because of load sharing between the plate and the bone. If a screw hole is empty, it becomes an area of stress concentration and is susceptible to increased mechanical cycling and early fatigue. The gliding hole is the basic component of the DCP.

a) 4.5 mm DCP narrow or broad -used in large breeds of dog

b) 3.5 mm DCP narrow or broad-used in medium to large breeds of dog

c) 2.7 mm DCP-used foe small breed of dogs

d) **Semitubular plate** (1/2, 1/3, 1.4 segments) The quarter tubular plate is the most useful, takes 2.7 mm screws, used in small breeds of dogs , and also in acetubular and mandibular fractures

e) **Small fragments plate**- Variety of shapes: "T" plates, angled plates used with 2.7 mm cortical screws in mandibular and pelvic fractures

f) **Mini plates**- can be used with 2mm or 1.5mm cortical screws, used in radius fracture, ulna fracture in toy and miniature breeds of dogs.

g) **Special plate**- the reconstruction plate can be cut to length and contoured in any direction. it is especially useful in pelvic fracture

h) **Double hook plates**- are designed to stabilize fractures and osteotomies near the ends of long bone.

i) **Reconstruction plate**- has oval holes, an equal distance between all holes, and a groove cut from the sides of the plate. This plates work well in areas where plate contouring can be difficult (eg, acetabular fractures). These plates are less resistant to bending forces. It can be bent and twisted into two dimensions. But it should not be bent more than 15^0. Decrease stiffness than DCP. Reconstruction plate is used where the exact and complex contouring is required: pelvis, distal humerus and cervical.

j) **Limited contact DCP (LC-DCP)** - have a grooved undersurface for less plate-to-bone contact, this feature allows for improved vascularisation of the bone. Though it has grooved undersurface, the LC-DCP has strength similar to that of DCP when tested by four-point bending. The LC-DCP is further development of the DCP is used for the same indication as DCP, but the improved design address additional advantages. The evenly distribute undercuts reduces the contact area between bone and plate to a minimum. This significantly reduces impairment of the blood supply of the underlying cortical bone undercuts and also allow for the formation of mini callus bridge. The LC-DCP is designed with the spherical gliding principle implemented at each end of the hole to ensure compression in either longitudinal direction. This becomes beneficial for segmental fractures.

k) **Locking Compression Plates (LCP)** - Broad and narrow, intended for fracture fixation of different long bones includes femur, humerus and tibia. It is used in fixation of periprosthetic fractures, osteoporotic fracture, multifragmentory fracture, non-union and malunion. LCP also can be used in MIPO or in less invasive technique because precise prebending of the plate is not necessary and no primary loss of reduction. The basic principles of LCP are its angular stability, whereas stability of conventional plate relies on the friction between bone and plate. The principle of LCP is screw locking. LCP provides the relative stability. In this fixation, fractured healed by secondary healing- callus formation. The mechanical properties of locked screw plate are plate sits slightly on the bone. Tightening the screw locks the screw head within the plate. The plate is not drawn towards the bone and there is no compression between the bone and plate. Screw locked the plate forming a fixed-angle construct. Once the locking screw engage the plate, no further tightening is possible. Therefore, implants locks the bone segment in their relative position regardless of degree of reduction. The main advantage of LCP is, it provides angular stability, axial stability, less damage to blood supply of the fractured bone, less soft tissue damage, does not need plate contouring before its use, chance of postoperative infection is minimum because it is a submuscular technique, reduced compression of the periosteum, protect blood supply the bone and bone healing by callus formation under the plate.

The prime indication for compression plate fixation is a simple transverse diaphyseal fracture of femur, radius and ulna. DCP, although means dynamic compression plate, it is not always used as a compression plate, it may also function as a neutralization plate or as a buttress plate.

Neutralization plate: any plate which is used without compression (especially for comminuted fractures). A neutralization plate acts as a "bridge". It transmits various forces from one end of the bone to the other, bypassing the area of fracture. It main function is to act as a mechanical link between the healthy segments of bone above and below the fracture. This plate does not produce any compression at the fracture site. A neutralization plate is applied after lag screws have been used to reconstruct the shaft in an oblique, butterfly fragments or a mildly comminuted fracture of a long bone or the fixation of a segmental bone defect with a combination of bone graft. The plate is used to protect this repair and transmits all stresses from the proximal to the distal end of the bone.

Buttress plate: A plate which is used to span an area of comminutuion where the fragments are too small for lag screw fixation. The mechanical function of this plate, as the name suggests, is to strengthen (buttress) a weakened area of cortex. The plate prevents the bone from collapsing during the healing process. A buttress plate applied the force to the bone which is perpendicular (normal) to the flat surface of the plate. The fixation of bone should begin from middle of the bone, i.e. the closest of the fracture site on the shaft. The screw should then be applied in an orderly fashion, one after the other, towards both ends of the bone.

Compression plate: A compression plate produces a locking force across a fracture site to which it is applied. The plate is attached to a bone fragment. It is then pulled across the fracture site by a device, producing tension in the plate. As a reaction to this tension, compression is produced at the fracture site across which the plate is fixed with the screw. The direction of the compression force is parallel to the plate. The effects occurs according to the Newton's third law (action and reaction are equal opposite).

Principles of plate fixation

Plates should be placed on the tension side of the bone (laterally in the femur; cranially in the radius; medially in the tibia; and medially, laterally or cranially in the humerus). A minimum six cortices proximal and distal to the fracture site should be engaged (3 screws on each side). Bone plating is not limited by fracture configuration, comminutuion or location. Before plating, contour the plate to the normal shape of the bone. The plate can serve as a compression, neutralization or buttress. Plate and screws should be made to same metal (except in horn plate) to prevent corrosion effect. Bone plates that are used for dynamic compression should be slightly prestressed or over bent. Plates that are not prestressed create only 20% bone compression which is limited to the near cortex. Prestressing a plate results in a 1-2 mm gap at the near cortex located under the plate but results in compression on the far cortex, thus increasing overall stability. A gap in the cortex under the plate allows the plate to resist tension forces, a gap in the far cortex results in the plate having to resist more bending forces. Because of their rectangular block design, bone plates are much more effective at resisting tension than bending forces.

Limitations: Requires complete exposure of the fracture site which greatly disrupted soft tissue attachment and bold supply to the fractured bone. When a large bone plate very rigidly fixed, it may excessively protect the bone from loading stresses. Since the bone requires some stress to stimulate healing, this "stress protection" can impede heal and weaken the bone. When a bone plate is applied distally on the extremities where little soft tissue coverage is present, the patient may become lame in cold weather. Implant failure, compartmental syndrome, sciatic nerve trapement, interference with tendon motion and infections are some other limitations of bone plating which is commonly encountered in small animal surgery.

(D) Tension band wiring

Tension band wiring is indicated for the treatment of avulsion fractures of the olecranon, greater trochanter, patella, tibial tuberosity and os calcis. In all these fractures, the fragment is distracted by the muscle, tendon or ligament which inserts on it. The tension band is places so that it counteracts the tensile force acting on the fragment and redirects it to compress the fragment against the adjacent bone. Tension band fixations are composed of dual K-wires that fix the avulsed fragment back to the shaft and wire in a figure-eight form that passes through a hole in the bone on the diaphyseal side of the fracture line. The purpose of k-wires is to maintain fracture alignment and provide some rotational stability. The principle of tension band fixation is that the figure-eight cerclage counteracts the tensile forces of the tension attachment on the avulsed piece and converts these forces into a vectored force that compress the fracture as the animal bear's weight.

(E) Interfragmental compression

Interfragmental compression is a method of compressing two fragments of bone together and is achieved by the lag screw principal. Any screw can be used as a lag screw. The lag screw is the most effective way to achieve compression between the bone and fragments. It pulls the fragments together producing pressure across the fracture line. The lag effect is achieved by overdriving the screw hole in the near fragment so that the screw thread grips in the far fragment only, then when the screw is tightened the bone fragments are drawn together and the fracture line is compressed. It is providing purchase on the distal fragments while being able to turn freely in the proximal. The screw must glide freely through the near cortex and engage only the far cortex/fragments. Whenever a screw crosses the fracture line, it should be inserted as lag screw. The goal of a lag screw is to compress the fragments together into normal anatomical alignment and to essentially eliminate gap and strain between the fragments. This encourages load sharing and load transfer between the bony fragments. Maximum interfragmentary compression is achieved when the lag screw is placed perpendicular across a fracture line.

Three types of ASIF screws

Cortical screw- Threaded in the whole length (fully threaded). Used in the hard cortical bone of the diaphysis.

Cancellous screw- Partially threaded, used in the soft bone of the metaphysis and epiphysis. Locking head screws (LHS): self drilling and self tapping

Biomechanics of plate fixation with conventional screw: Screws in tension, plate/ bone friction, compression at fracture site and disturbed blood supply;

Biomechanics of locking head screw (LHS): Screw in shear, noncontact plate, no compression of fracture and preserve blood supply. Function of LHS: always in combination with plate, never as lag screw, never cross an unreduced fracture with a LHS. Drawback: Screw insertion is only possible in a 90^0 angle and screw jumping and difficult implant removal.

Screw length is divided into head, shaft, thread, and total screw lengths. It is important to understand that total screw length is measured when using a depth gauge to determine the screw length required for the hole drilled. Screw size refers to the thread diameter of the screw. Thread diameter is determined by the outside diameter of the thread. The internal diameter of the screw is the diameter of the shaft at the threaded part of the screw.

8. ASIF principles of fracture repair

ASIF (Association for the Study of Internal Fixation) principal of fracture repair is to restore full function to the injured limb as quickly as possible. This can be achieved by

- Atraumatic surgical technique
- Accurate anatomical reduction
- Rigid internal fixation
- Avoidance of soft tissue damage

DCP

LC-DCP

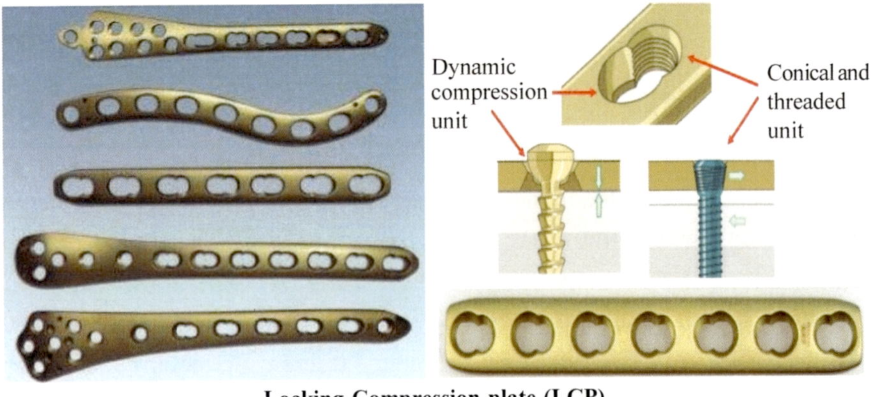

Locking Compression plate (LCP)

Conventional screws
Cortex screw (with or without shaft)
Cancellous bone screw (with or without shaft)

Locking head screws (LHS)
Self-tapping LHS
Self-drilling, self-tapping LHS

Different types of Screws

[Acknowledge for reproduce from Slide share of Michael Wagner (2012), Ahmad Fadzli Sulong (2017)]

- Rigid fixation can be achieved by compression techniques, includes- Functional compression as in tension band wiring; Interfragmental compression achieved with lag screws; Axial compression, achieved with a plate or a wire using the tension band principle; Interfragmental compression used in combination with a neutralization plate.

9. Fracture healing time and removal of fixation devices

Time to reach clinical union

Age of animal	Intramedullary pin/plate fixation
Under 3 months	2-3 weeks/4weeks
3-6 months	4-6 weeks/2-3 months
6-12 months	5-8 weeks/3-5 months
Over 1 yr	7-12 weeks/5months -1 year

The above timings may vary with the animal to animal, age, status of fixation, asepsis, postoperative care and nutrition, type and site of fracture, bone involved and extent of soft tissue damage.

10. Post operative care

Successful fracture repair does not end when the last suture has been placed. Specific areas to be address which includes exercise, restrictions, cage confinement, physical therapy, and medications, signs of complications and conditions and limits for follow-up examinations. Uses of broad-spectrum antibiotics along with non-steroid anti-inflammatory drug (NSAID) are essential for at least 7-10 days depending upon the fracture condition. Suture/Incision site should be dressed daily with some mild antiseptic solution eg. Povidone- iodine and some fly-repellent antibacterial cream should be pasted on the suture line until it healed properly. Good diet enriched with calcium, phosphorus and vitamin D and B-complex should be provided to the animal to ensure rapid restoration of health and accelerate fracture healing. Early use of the affected limb and client education are also needed for more success.

In summary, veterinarians must be familiar with different types of fracture fixation techniques, advantages and disadvantages of these techniques, thorough examination of the patient, proper diagnosis and careful evaluation of each fracture, chances of complication including operative and postoperative infection, successful rate, availability of fixation devices, cost of these devices, cost of the total operation including anesthesia and postoperative treatments, cost of the animal, ethical issues and finally the economic status of the animal owner. It must be remembered always that no two fractures are the same, thus very careful fracture evaluation and preoperative planning is utmost necessary for any successful fracture treatment.

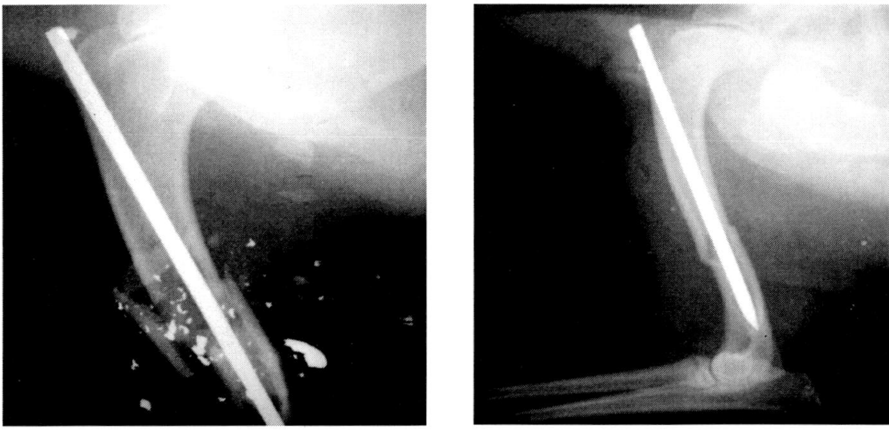

IM Pinning

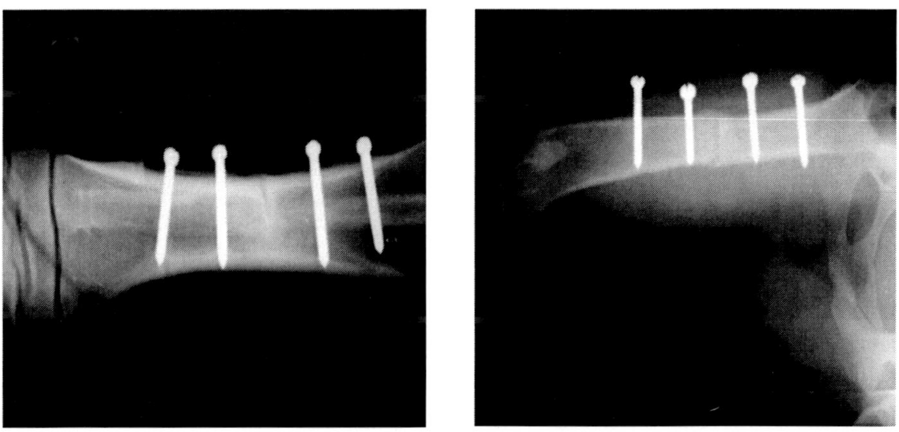

Horn Plating

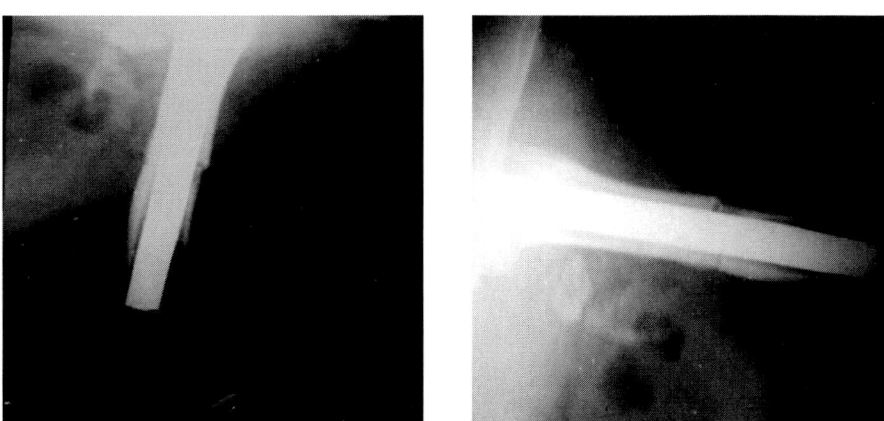

K-Nailing

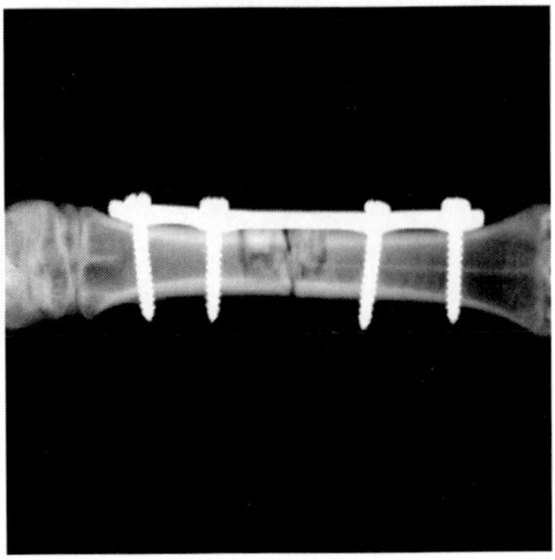

Bone plating with bone graft

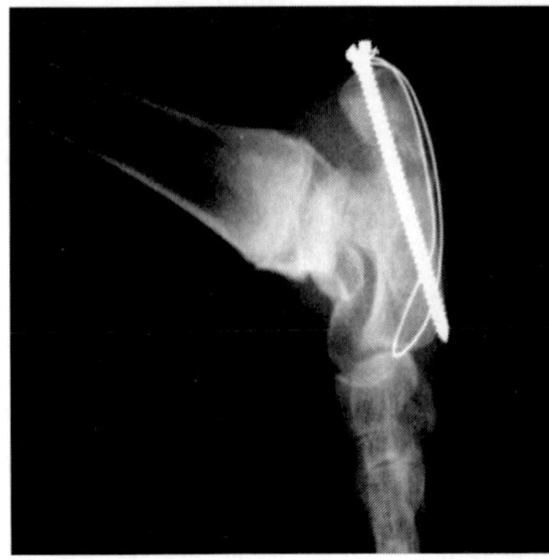

Tension band wiring

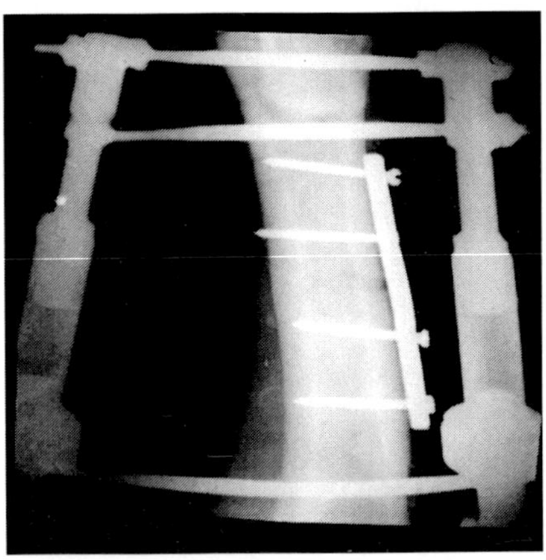

Internal & External Fixation

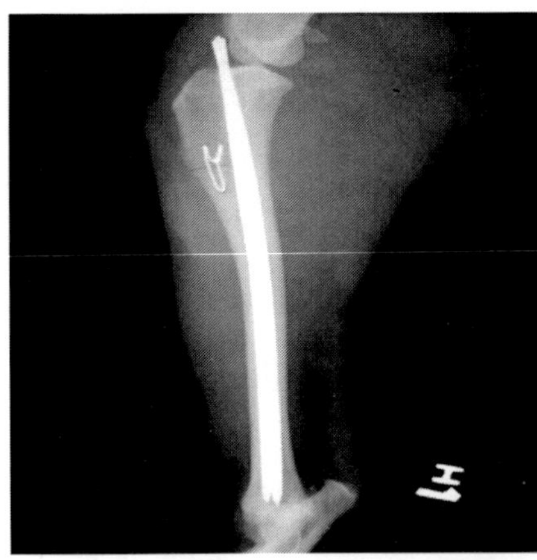

Cross pining + interfragmentary wiring

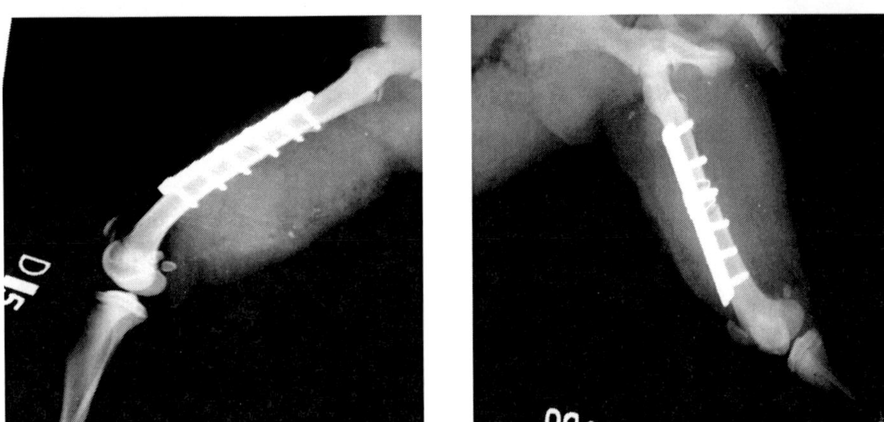

Diaphyseal comminuted #, DCP with full cerclage wiring

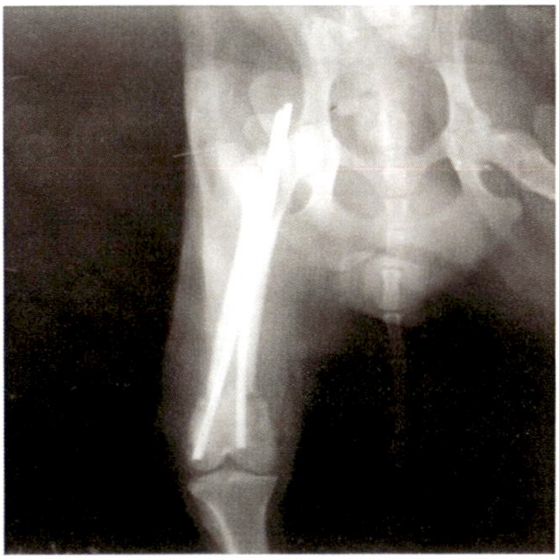
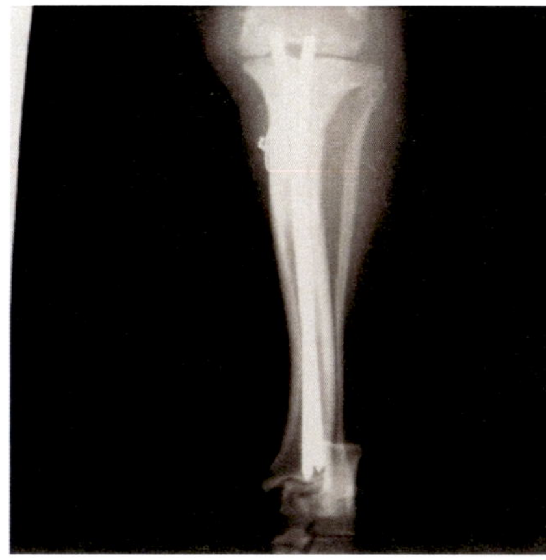

Supracondylar fracture treated with cross pinning

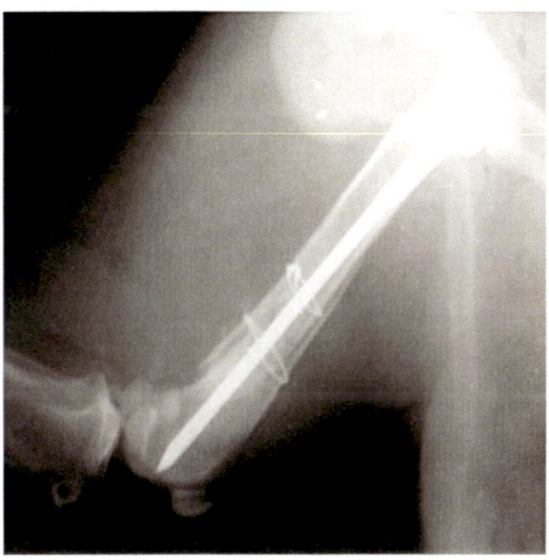

IM pining with full cerclage wiring

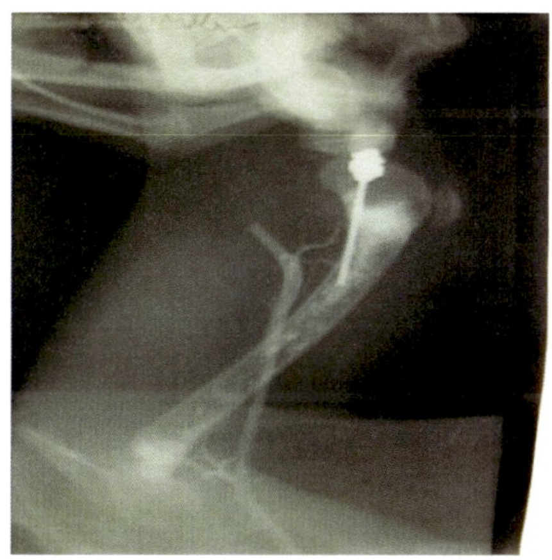

Osteomedulography

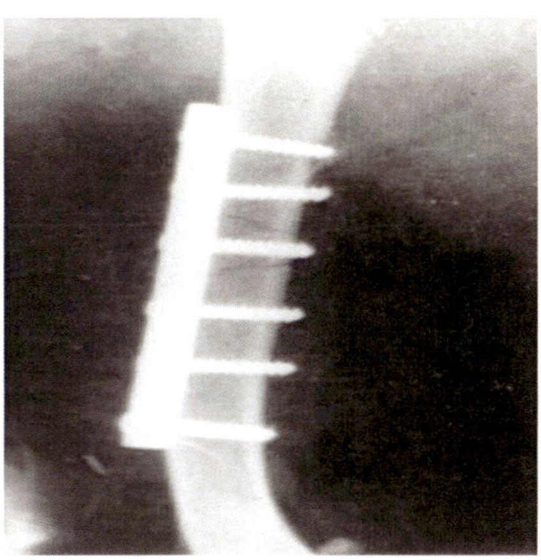

Communicated # with DCP

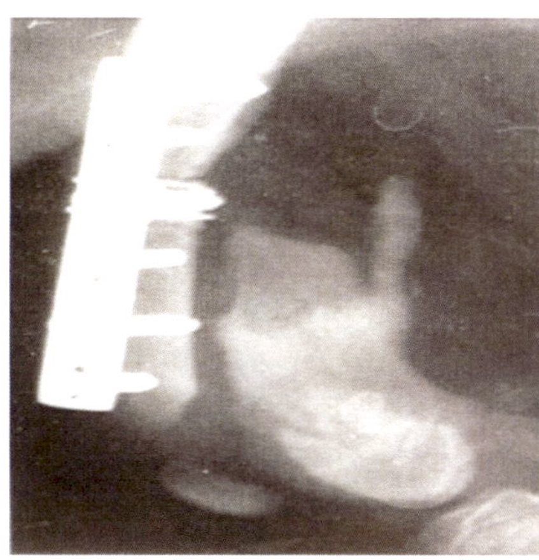

Complication of Internal fixation

6

Application of Plaster Cast/Splint for Fracture Immobilization and Other Bandaging Procedure in Small Animals

Fracture in animals occur frequently and is a concern of economic importance to the owner. The practice of keeping them as pets in households rather than in sheds or in confinements have increased the chance of encountering accidents like fall from coat, terrace, limbs being trapped in drains, being hit by vehicles, dog bites etc. leading to fracture. Fracture management in small animals/small ruminants can be successfully achieved owing to their light weight; they tolerate external coaptation relatively well and can ambulate in three limbs well in order to spare the injured limb. Fracture can be repaired successfully with regional anaesthetic techniques which they can tolerate very well. The constraint of cost effectiveness can be balanced by carefully selecting the appropriate technique according to the weight and limb involved. Many economical skeletal fixation technique have been evolved which can well fit into the budget of the owner. Knowledge of simple techniques can enhance healing in cases of bone loss and management of infection.

Fracture site stabilization

Fracture repair is usually elective, hence initial support and care of the fracture is necessary to prevent further trauma and complications like communication of fracture site to the exterior through wound or sinus and chances of infection. Initial support to splint the bone not only limits the movements but also supports the soft tissues from untoward consequences. Limb splinting reduces edema and pain, preserves neurovascular supply to the limb till the day of operation and helps to transport the animal safely. Fracture to limbs caudal to mid radius ulna/tibia can safely be stabilized by splints or full limb casts. Wood, pliable bamboo or split poly vinyl chloride pipes positioned on the caudal and lateral aspects of the limbs and supported firmly with tapes or bandages can serve the purpose. Fracture proximal to mid radius ulna or tibia should not be stabilized using splits since the fulcrum effect created due to this can augment soft tissue injury.

Careful assessment of the viability of the structures caudal to fracture site should be assessed prior to a decision regarding surgery, devitalized tissue hamper healing and is a subject to be considered for amputation.

Signs of fracture

The history of improper limb carriage or non weight bearing, associated with a traumatic event or as a sudden finding can implicate a possibility of fractured limb. Pain and acute swelling may hinder palpation of the site, though most animals are stoic without much pain and shrieks. Discontinuity, crepitus and abnormal alignment of the bone with tenderness can be appreciated. Radiography should be undertaken with orthogonal views (anterio-posterior and lateral) unless in specific situations can elaborate the nature of fracture, enough information to plan and execute fracture repair. Good technique and proper positioning is imperative to detect even subtle abnormalities like hair line, the involvement of joint or the level of comminutuion. Sedation with butorphanol (0.05 mg/kg IV), Xylazine (0.05-0.1 mg/kg im) or Midazolam (0.2 mg /kg iv) can give adequate muscle relaxation if required for taking radiographs.

Considerations for fracture management

The position of fracture which is either open or closed, the level of soft tissue damage and the viability of limb distal to fracture, suitability of implants experience of the surgeon and cost effectiveness of techniques are to be considered and the technique should be planned.

- Location and configuration of fracture
- Biomechanics of fracture
- #Forces: Tensile, compressive, bending, shear & torsion
- Concomitant orthopedic and soft tissue injury
- Age, size and temperament of the animal
- Pre-existing disease
- Overall health status of the animal
- Availability of the instrument/devices
- Expected performance level of the animal
- Economics

Anaesthesia for surgery

Regional analgesic techniques are better suited to small /large ruminant fracture repair owing to the safety and simplicity of the technique and lesser peri- operative complications. Intravenous regional analgesia (IVRA) technique is popular in this regard. For forelimb surgeries, a rubber tourniquet is approximately 1-1.5 cm wide and 10-15 cm length can be used. The forelimb to be operated is restrained with the same side lateral recumbency, and the bandage is applied cranial to the elbow joint, the limb caudal to this is aseptically prepared for surgery. The cephalic vein is located on the medial middle third of the forearm, 3-4 ml 2% lidocaine is injected into the vein. For hind limb surgery, the limb considered is to be restrained up, opposite to the site of recumbency, the bandage is applied above the hock joint and limb caudal to this site is aseptically prepared. The recurrent tarsal vein is located at the groove in front of the gastrocnemius tendon and 3-4 ml 2% xylocaine is injected into this. The analgesia will remain for 30-45 min after which the tourniquet can be released. For surgery involving the femur or hind limb, lumbar epidural /spinal analgesia with lidocaine (0.2-0.4 mg/kg b.wt) or Xylazine- ketamine (0.05 mg/kg and 2.5 mg/kg, 0.75 ml each) is suitable. General anaesthesia or balanced anaesthesia is often employed for surgery involving humerus or femur when the technique is time consuming.

Fracture reduction

Reduction of a fracture requires general anaesthesia and good muscle relaxation. Suitable fractures for external fixation are transverse or short oblique fractures, which are stable when, reduced. If the fracture can be reduced but slips easily out of alignment, it is not suitable for external fixation. Reduction depends on traction to reduce muscle spasm and angulations of the fracture tend to lever them together. Great care is necessary not to convert a closed fracture into a compound fracture by breaking the skin. If a fracture proves difficult to reduce, an open reduction and internal fixation should be carried out rather than causing extensive soft tissue damage.

Initial step for fracture treatment

During evaluation of an animal with a fracture, first assess the patients overall health. Trauma severe enough to result in broken bones can also cause life-threaten injuries such as pulmonary injury or hemorrhage into the abdomen. During the initial stages, cover any open wounds over the fracture site with sterile dressings and temporarily immobilize the fracture. Sterile wound coverage reduces the likelihood of nosocomial infections that may complicate wound and fracture healing ultimately. Temporary fracture immobilization prevents further soft tissue injury and makes animal more comfortable.

Guidelines for management of open fracture

Highly comminuted fractures are typically associated with severe soft tissue injury and it is more commonly found in animals. These tissue injuries must be treated in conjunction with fracture. Delays in fracture healing, infection and other complications can arise if the surrounding soft tissue not properly treated. Open fractures are typically treated as emergencies. In general, the more extensive bone and soft tissue damage, the greater the chance of complications and it need immediate intervention. Parenteral antibiotic therapy should be started immediately along with debridement of all dead tissue. After debridement, rigidly stabilize the fracture to optimize fracture and soft tissue healing. Inadequate fracture immobilization not only leads to delay in fracture healing but also slows the repair of surrounding soft tissue injury.

Any fixation device can be used for the stabilization of an open fracture which is appropriate for a closed fracture of the same configuration. Properly applied different internal fixation devices such as intramedullary pins with cerclage wiring, interlocking nail and bone plating, provides adequate stability of the fractured bones and simultaneously they do not substantially diminish blood supply to the fracture site. The concept is that using intramedullary (IM) devices in open fractures is inappropriate, but in my clinical experience it does not pose a critical problem when combined with proper debridement and wound care. Final resolution of osteomylitis may require removing of implants when fracture healing is complete. It is important to note that a bone will heal, even if infected, if the bone is rigidly immobilized and stabilized.

Principles of some common external fixation methods

Splits

- First aid measure for temporary immobilization of fractures of lower long bones
- Easy to apply, offer good strength
- Woods, PVC pipes, metal bars
- Thermoplastic splints-large animals

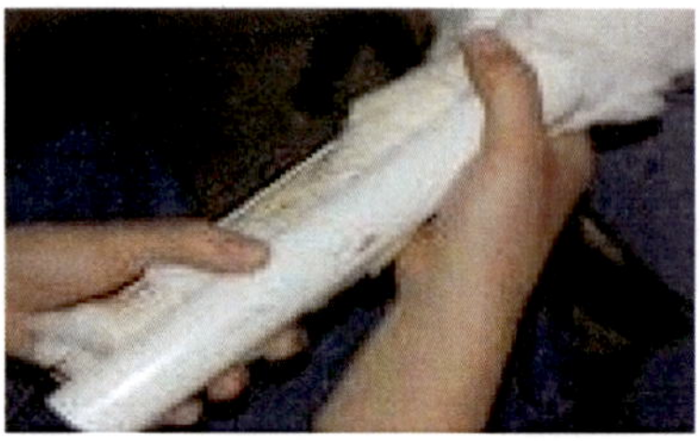

Application of splints

Application

- A thick bandage is applied for padding.
- PVC splint is held in position using elastic leg wraps.
- Adhesive tape is applied to top and bottom to avoid slippage of splint.
- Avoid edema.
- Should remain in position for 2 to 3 weeks or until animal is recovered

Limb casting

The casting material according to the availability and economic concerns can vary. Plaster of Paris cast is easily available and economical. POP casts can be made locally or readymade material can be purchased. PMMA (Poly Methyl Methacrylate) or fibre glass are other choices. Fibre glass casts have advantages of strength and stiffness; do not weaken by moisture and stress. Fiber glass are stronger, lighter in weight and radiolucent in comparison with plaster of Paris, however, they are more costly. Adequate cotton padding or foam resin to cover bony prominence, put in place with the help of bandages or stockinet precedes cast application. POP cast is adequate to manage closed, nondisplaced fractures of caudal radius ulna, metacarpals or metatarsals without comminutuion or sharp ends. This can be put in place for 15 days at a stretch and evaluate the strength, and reapply the cast if required. Bamboo splints or PVC strips can reinforce the strength of the cast when incorporated with the initial padding. Care must be taken that the cast is applied not too tight to obstruct the venous return which can lead to swelling of the toes.

The casts can be applied as either full limb casts or half limb casts according to the bone to be immobilized. Half limb casts can be used to immobilize closed fractures of metacarpal or metatarsal bone, un-displaced physeal fractures and fractures involving phalanges. Cast application ensures the immobilization of one limb above and below the bone fractured. Comminuted fractures and fractures involving the articular surface require rigid fixation techniques where cast application is uncalled for. Full limb casts can be applied for closed transverse fracture of distal radius ulna or proximal metacarpal or metatarsals. The time for fracture healing under casts depends upon the stability of the cast, age of the animal, presence or absence of infection and fracture stability. An average 20 to 30 days time is sufficient for the retention of cast. Cast complications like sudden onset of pain, swelling or sores alerts cast removal.

Plaster Cast

Indications: External immobilization if the fractures below the elbow and stifle joints, ligamentous injuries

Material Required: Plaster of Paris (Gypsum), Cotton, Bandage and different types of Splint (bamboo, wooden, PVC strips), fiber cast.

Technique

- Animal is restrained under sedation or general anaesthesia on lateral recumbency with affected limb uppermost.
- Fracture fragments are reduced by traction or counter traction to their original position
- Zinc oxide or talcum powder sprinkle over the limb

- Layer of non absorbed cotton is applied around the leg to protect bony prominces. Over or less padding should be avoided which causes impair immobilization
- Splints are bandaged and placed either in latero-medial or anterio-posterior side of the limb and tied with fractured limb.
- Plaster of Paris bandages are soaked in luke-warm water for few minutes till air bubbled stop
- Plaster of Paris bandage are then removed from the water, squeezed gently and then wrapped over the limbs, includes splint, staring at the fracture site and continuing up to down including the entire limb, except the toe in dogs and hooves in large animals.
- Plaster cast wrapping over the fractured limb should not be too tight or too light, which causes improper pressure on the limb leads to edema or loosening of the cast.
- After completion of wrapping, it should be molded by gentle rubbing with wet hand to give a smooth and uniform shape.
- Then it should be allowed for dry till its well setting and become hard.

Few points should be remembered during Plaster Cast application:

- Joints above and below the fracture site should be immobilized with plaster cast.
- No movement should be allowed during cast setting
- If edema or swelling observed either below or above the cast, a change of plaster cast is necessary after subsides the edema.
- In case of open fracture, a window should be cut in the cast at the level of skin wound for drainage and daily dressing of the wound.
- Radiographs should be taken at different intervals to examine the alignment of fractured bone and formation of callus.

Limitations

- Cannot be used in humerus and femur fracture
- Cannot be done in long oblique and comminuted fracture

Thomas splint

Thomas splint can be useful for fractures at proximal radius ulna or tibia, caudal to humerus or femur. The round frame and the hoop of the Thomas Splint help in additional anchorage and weight transfer into the frame from the body, thus sparing the fracture site. The complications involve inadequate fixation of the frame with the limb leading to rotation of the frame and displacement of the fracture fragments. Adequate space should be provided while the initial frame built in order to incorporate cotton padding. The hoop of the frame should support the axilla or inguinal area for support with body wall and weight transfer into the frame. The animal may need to accustom with this before proper weight bearing.

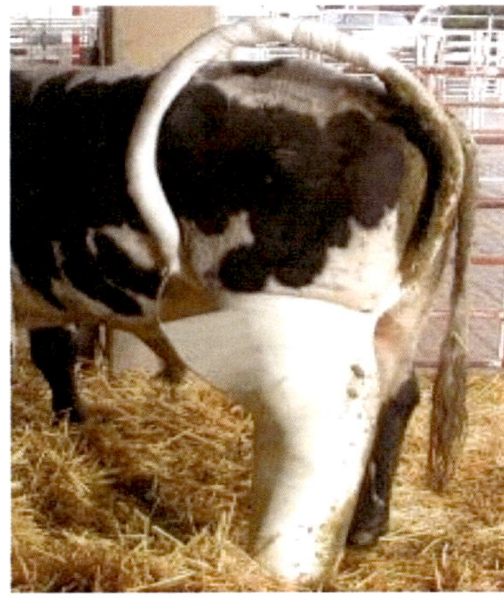

Thomas splint application in a cow

Materials required: Aluminium rod or conduit pipe of different sizes, cotton, gauze bandages, adhesive tapes, splint mold, sedative or tranquilizers.

Methods

- The frame of the splint is made of aluminum rods available in a variety of diameter.
- The first important component to manufacture for the splint is the upper ring, which fits comfortably into axillary or groin areas
- The ring should be oval and bent towards the animal body
- Ring diameter – from tuber ischi to tuber coxae <H L>
- From axilla to mid point of scapula<F L>
- Total length of the rod is calculated by the formula:-2(3D+1)+2L+20

  D-diameter of the ring

  L-Length of the splint bar
- For fore limb- the distance from the axilla to a mid-point on the scapula will be the diameter of the ring. The length of the bar will be the distance between the axilla and the tip of toe
- To prepare the ring, a splint mold is used and the rod is bent around the splint mold
- The extension rods for the fore limb should be straight but in the hind limb they should be bent to conform to the natural angle of the stifle and hock

- In the splint for bovine, an oval shaped iron sheet of the size of hoof is welded to provide a suitable anchorage for the hoof. For the dog, splint the lower ends of the bars should be bent and overlapping ends are welded.

Technique

Under deep sedation/tranquilization the animal is restrained in lateral recumbency with the fractured leg uppermost.

- Ring and bars are well padded with cotton and gauze bandage.
- Overriding fracture fragments are reduced by traction or counter traction using a cotton rope tied at the level of toe/hooves
- Entire limb should be padded with cotton bandage
- Splint is applied on the limb and fixed with the help of adhesive tape or bandage. Based on the location of fracture and displacement if bone fragments, the traction is applied at different levels and the bones are fixed to the frame using gauge/bandage.

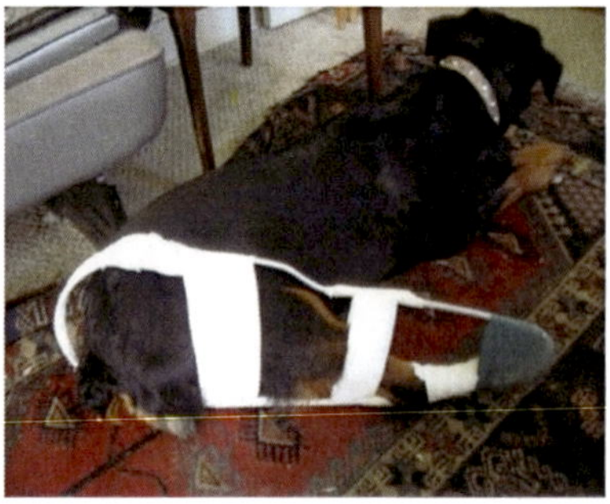

Thomas splint application in a dog

Limitations

- Not applicable in long oblique or comminuted fractures.
- Fractures near the joints.

Complications

- Abrasions and rubbing sores in axillary and groin regions
- Prolonged use-disuse atrophy of muscles, osteoporosis of bones, pathological fractures.

Hanging limb pin cast

A hanging limb pin cast refers to the placement of transcortical pins through the proximal fracture fragment and incorporating the pins into a cast. The cast begins proximal to the pin application, incorporating the transcortical pins and the entire limb caudal to it including the foot. The forces acting on the bone spares the fracture site, gets transmitted through the pins and the cast into the ground. This can be helpful to treat comminuted metacarpal or metacarpal fractures by biological osteosynthesis since it causes minimal fracture site disruption. This technique is cheaper and effective once the size and placement of the pins are appropriate. Care must be taken to spare the joints, articular surfaces and growth plate in young animals.

The diameter of the pins should not exceed 20% of the diameter of the concerned bone; the stoutest pin should be placed most cranially since it takes away majority of the forces acting on the bone. Pins should spare at least six pin diameter space from the fracture ends and the individual pins placed at four pin diameter spacing. Align the pins at a angle of 30^0 to provide rotational stability of the construct. The cast material should be applied closer to the skin after bandaging. The problem of poor access to wound at the area is a disadvantage, the windows if left for wound care should not weaken the strength of the cast. Look for window edema which can be managed by bandaging the wound at window and regular check for swelling and infection.

Modified coaptation splint

- Hind limb-for management of tibial fractures
- Also used to stabilize stifle joint
- Prevents medial displacement of fractured tibia
- Coaptation splint assembly-3 parts: crutch, leg bars & foot plate

Walking cast

Indication

- Fractures of distal radius, tibia, metacarpus, metatarsus, phalanges
- Comminuted fractures that can not be adopted by other techniques
- Where it is necessary to protect the digit from weight bearing

Principal

Two or three Steinmann pins passed transversely through proximal fragment and pins are connected to U shaped steel/ aluminum or iron frame and incorporated in a plaster cast, allow the animal to rest on the frame and thus relieve the fractured bone from direct weight bearing.

Procedure

- Animal is restrained in lateral recumbency under caudal epidural or general anaesthesia.
- A steel wire/rope is fixed to the hoof wall and traction applied to the limb to facilitate fracture reduction.

- After giving a stab incision in the skin, 2-3 pins are placed through the bone proximal to the fracture site.
- Wounds around the pin hole are closed with synthetic absorbable suture material and covered with sterile gauge.
- About 3-4mm thick POP is applied over sterile layer of cotton wool and gauge from hoof up to the joint above the fracture site.
- The aluminium/steel/iron frame is then modulated to the shape of the casted leg and placed on the pins after drilling holes in the frame at the level of pins (the holes in the frame should be 3mm larger than the pin diameter to prevent torsion between pin and frame).
- The pins should be cut in such a way that both ends will protrude approximately 2.5 cm outside the steel frame.
- The space between the pins and frame is filled with methylmethacrylate (Techonovit) to hasten the frame to the pins. The protruding ends of the pins are also covered with Techonovit to protect the animal as well as the pins.
- A second layer of plaster is then applied incorporating the frame, which also helps to fix it to the first layer.
- The cast and frame should reach up to the joint above the fracture site, to prevent secondary fractures at the site of the proximal pin. The distal part of the frame is not covered by the cast and the hoof is not fixed to frame.
- After hardening the cast, the traction wire/rope should be cut.

Postoperative care

- The patient should be confined to the stall, provide rest during the healing period.
- The pin sites should be watched regularly for any osteolytic changes, infection.
- The cast should remove after x-ray examination confirmed complete bone bridging.
- After removal of pins, skin should be dressed regularly till healed.

Figure Eight (EHMER) Sling

- Synonyme:-figure 8, butterfly, ehmer sling.
- Designed to create flexion of the rear limb with rotation of the femoral head into the socket

- Indication: Provide stability following reduction of cranial and dorsal dislocation of hip joint

Methods

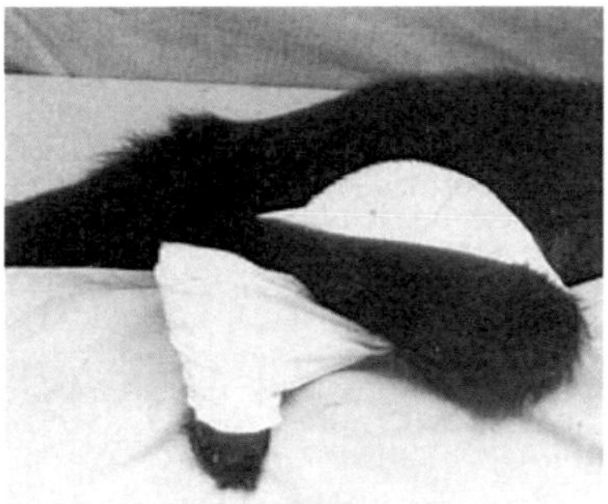
Ehmer Sling application in a dog

- Adhesive tape sling is made for the metatarsal area.
- Bandage is passed on the medial side of the thigh to the proximal flank and is returned to the point of origin.
- Passed on the medial side of the leg to the flank again
- Bandage is placed on the lateral thigh and medial hock.
- Continued ventral and lateral to the metatarsal and again to medial aspect of thigh.
- Foot rotated inward and hock outward.
- An accessory tape encircling the sling, abdomen, and back in cats and short legged dogs for additional stability.

Velpeau Sling Bandage

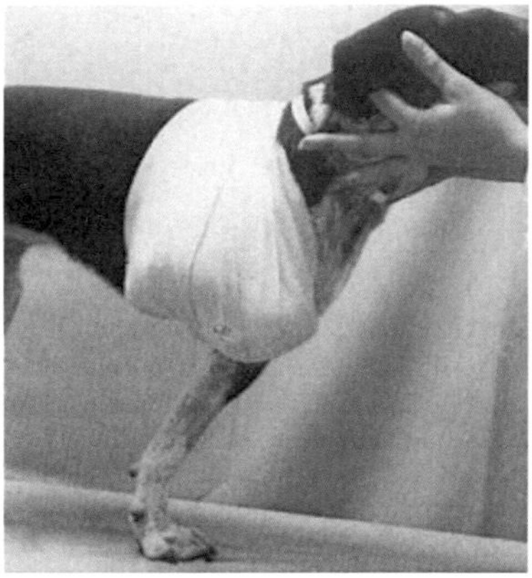
Velpeau sling is in place

- The Velpeau sling is a shoulder bandage that will relieve the forelimb of weight bearing.
- It keeps the carpus, elbow, and shoulder joints in a flexed position.
- Used as comfortable sling for the forelimb.

Indications

Immobilization of scapular fractures, dislocations of the shoulder joint, for the removal of weight bearing from a fore leg.

Methods

- A conforming gauze bandage is wrapped around the paw, the leg is flexed and the bandage is brought up over the lateral aspect of the shoulder and around the chest.

- Several layers are applied and then covered with elastoplasts.
- Taped to the body wall in a way that prevents the elbow from coming out posterior and the forepaw from coming out interiorly.

Robert Jones Dressing

- The Robert Jones bandage is a thick cotton wool bandage which acts as a splint and controls edema
- It is useful not only as a first aid measure for the temporary immobilization of fractures but also as a postoperative bandage for fractures which have been treated surgically.
- The bandage is comfortable to wear and is generally well tolerated despite its bulk
- Bulky bandage for immobilization of an extremity
- Immobilizing joints following reconstructive surgery, such as cruciate or patella surgery
- Large bulky dressing wrapped very tightly to the extremity, provide supreme comfort and relative immobility to the injured site
- Promote healing by immobilizing the injured area, thereby limiting swelling and providing protection from secondary trauma.

Transfixation pins and cast

Transcortical pins through both the proximal and distal fracture fragments and casting with either POP or fibre glass can be achieved in closed fractures of mid shaft radius ulna or tibia in larger ewes and goats. Here the cast will cover the bone conserved proximal and distal to the trasfixation pins. The amount of casting material is less and the animal can ambulate in its legs after its application.

External skeletal fixator (ESF)

- It is indicated for compound/ comminuted fractures of tibia, radius, and metatarsus, metacarpal.
- External skeletal fixation requires at least two transcortical pins placed proximal and distal to the fracture site and connected through connecting bars, fibre glass cast, epoxy putty, or PMMA.
- External skeletal fixation is versatile, easily applied, augments biological fracture repair and wound care. Pin size can be selected according to the size of the animal and bone involved.

- The fracture forces are shared equally by the individual pins. Fibre glass or epoxy putty more economic, easy to apply and are readily available in the market.
- The construct can be eventually disassembled to increase axial weight bearing.
- Postoperative care: Physical activity should be restricted and the animal should be confined to reduce the like hood of fixator entanglement.
- Pin-skin junction should be regularly monitored for any evidence of pin-track infection or sepsis.
- The fixator should be removed after radiographic evidence of complete healing.
- Main complication of ESF: Pin-track infection, implant failure, pressure necrosis of the skin, animal discomfort and skin damage.
- ESF should not be used where daily dressing or postoperative care is not possible.
- ESF should not be practiced where some simpler and less invasive method like external coaptation method is available, which would enable successful bone heal.

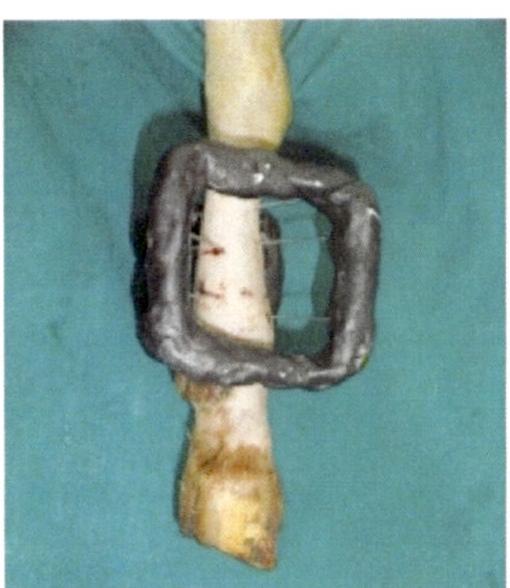

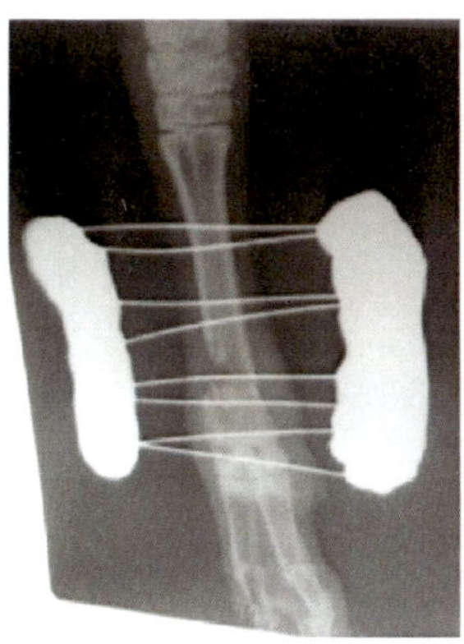

External skeletal fixator made up of Epoxy for metacarpal fracture repair.

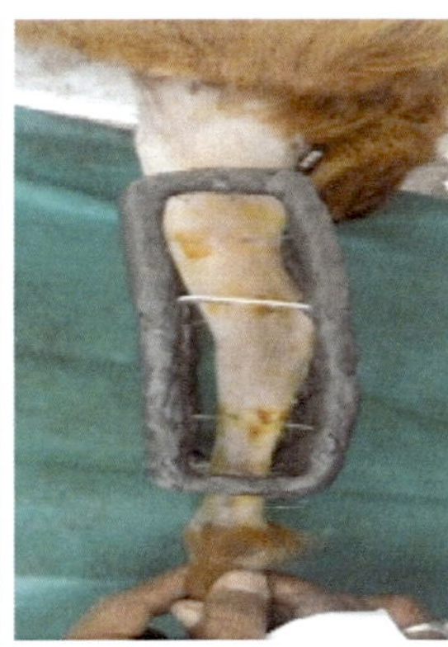

Managemnet of hock joint dislocation with External skeletal fixator (Epoxy)

Complications of fracture repair

Post operative immobilization/ limited ambulation is challenging in animals. Infection, delayed union, malunion, recurrence of fracture, implant migration etc. are frequently encountered in small animals. Adherence to proper aseptic protocol, minimally invasive technique and least surgical trauma and preserving maximum extraosseous blood supply to the fracture site can minimize such complications.

Conclusion

Fracture repair in small animals can pay very well to the efforts of the surgeon if conducted with appropriate implants and proper technique. Commonly available and economical implants and awareness about the techniques can render fracture repair in small animals a success and can prevent economic losses to the animal owner.

7

Bone Morphogenetic Proteins-Novel Regulators of Bone Formation

In 1965, Urist demonstrates that the implantation of demineralized bone at extra skeletal sites induced *de novo* formation of cartilage and bone and proposed that the decalcified bone matrix possess high osteoinductive property and the inductor is a unique chemical that differentiates mesenchymal type cells (pericytes) into new bone. Urist and Starates (1971) reported that the inductor is a protein and termed it bone morphogenetic protein (BMP) . BMP is the bone induction principle.

Sources: Bone morphogenetic protein isolate from bovine as well as human bone and designated as bBMP and hBMP respectively. The molecular cloning of the bone morphogenetic proteins and their subsequent expression in recombinant system, resulted recombinant bone morphogenetic protein designated as rhBMP. BMP has been extracted from dentin and bone matrix, as well as from bone tumours, by enzymatic and nonenzymatic methods. The sources of BMP, thus far explored are rabbit dentin, rat, rabbit and bovine bone, human bone, mouse or human osteosarcoma and human placenta.

BMP circulates in the blood and it can be detected by BMP-radioimmunoassay. BMP levels higher in growing children (20-72 ng/ml) and patients with Paget's disease (over 259 ng//ml) than in normal adults (14-18ng/ml). Crude and partially purified BMP has been extracted with the aid of collagenase, ethylene glycol, calcium chloride urea and Guanidine hydrochloride (GuHCL) solution. Wang et al (1988) purified BMP factor that induces bone formation> 300,000 fold from Guanidine chloride extracts of demineralized bone. As littlie as 50ng of this factor can induce the formation of cartilage and bone in an *in vivo* assay, though only 40μg could be obtained from 400kg of bovine bone powder.

Physiochemical property: BMP has not yet been completely purified but the following physiochemical properties of bBMP have been reported: (i) it is an acidic polypeptide, (ii) binds to hydroxyapatite, (iii) no carbohydrate detected, (iv) soluble in natural salt solution at pH 7.2, (v) degraded by acid alcohol solution, alkali sensitive, (vi) insolubilized by forming aggregates with Mr 14,000 proteins,

(vii) may contain one residue of hydroxyproline, (viii) insoluble in chloroform/ methanol, absolute alcohol and acetone, (ix) BMP Mr 14,000 complex, insoluble in Triton X100 (non-ionic detergent), insoluble in 0.6 M HCL, soluble in 6 M urea or 4 M Gu-HCL, (x) nuclease (RN-ase and DN-ase) resistant, (xi) Trypsin and chymotrypsin labile, (xii) resistant to chondroitinases A,B,C, amylase, neuroamidase, hyaluronidase, alkaline and acid phosphatase, chymopapain, collagenase, tyrosinase and thermolysin, (xiii) it has disulphide bond and it is inactivated by heat > 70^0C, deamination in HNO3, β-mercaptoethanol reduction, lathyrogens, penicillamine and ultrasound, (xiv) it retains its activity after proteolysis with pepsin or Trypsin and chymotrypsin, (xv) prolonged pulverization denatures /degrades BMP and (xv) BMP has an essential characteristic of a hydrophobic glycoprotein.

BMP is not species specific; rabbit BMP induces new bone formation in rat. Tyrosine is essential for the biological activity of BMP. Acetylation of tyrosyl groups blocks osteoinduction. BMP is soluble in body fluids and depending on the dose, is absorbed within a period of a few hrs or days; a single exposure did not incite any detectable cell-mediated immune response. BMP is normally present in the body fluids. It is therefore, not a foreign substance within the host. BMP is either covalently bond to the teleopeptide portion of the bone collagen molecule, or itself a teleopeptide, or part of teleopeptide protein complex of two. BMP contains as high as 36 cysteine and 6-methionine SH bearing amino acids which could react with IAA: iodoacetic acid SH-group enzyme inhibitor. High concentration of IAA degrades BMP.

BMP-ase: A neutral proteinase, BMP—ase, also a constituent of organic matrix of bone, degrades bone morphogenetic protein (BMP) within 24 hr after a bone is excised from the body. The extreme acidity and low temperature of the demineralizing solution (0.6 N HCL at 2°C) protects BMP against BMP-ase degradation. EDTA, pH 7.4, particularly at room temperature provides condition near maximal to BMPase activity and degradation of BMP.

Chemistry and types: The bone morphogenetic proteins (BMPs) are sub-family of polypeptide growth factors encoded by closely related genes-belonging to expanding transforming growth factor beta (TGF- β) super family. There are at least ten bone morphogenetic proteins (BMP—1 to BMP-10), butBMP-1 is not a member of transforming growth factor beta (TGF- β) super family.

Transforming growth factor beta (TGF- β) super family

(A) TGF- β subfamily

TGF- β_1

TGF- β_{2-} Found in many species, including humans

TGF- β_3

TGF- β_{4-} Found in chickens

TGF- β_5 Found in amphibians

(B) BMP subfamily

BMP-2 (also known as morphogens, 2A)

BMP-3 (also known as osteogenin)

BMP-4 (also known as BMP-2B)

BMP-5

BMP-6

BMP-7 (also known as osteogenic protein, OP-1)

BMP-8 (also known as osteogenic protein, OP2)

BMP-9

BMP-10

(C) Inhibin activin subfamily

Inhibin α

Inhibin βA

Inhibin βB

Mullerian inhibiting substances

Members of each subfamily shows close genetic sequence identity among themselves (50—90% sequence homology) and lesser identity with members of the other subfamilies (30-40% sequence homology). Members of bone morphogenetic protein family are somewhat more closely related to other distant transforming growth factor—beta like molecules than to authentic transforming growth factor-beta; within their own subgroup, they possess a number of similar physical characteristics.

Comparative amino acid sequence analysis of the mature proteins of BMPs has allowed furthering dividing these molecules into three distinct groups. Members of the first group, BMP2 and BMP-4, are very closely related (86% amino acid identity) to one another but their sequences differ significantly from those of BMP-3, 5, 6 and -7. BMP—5, BMP-6 and BMP—7 exhibited 71—80% identity to each other, thereby defining the second group. Of all the TGF—β like BMPs, BMP—3 is the most distinct and by itself forms the third group. BMP-5,-6 and -7 also share a high degree of identity in the propeptide region (52—64%). The propeptides of BMP-2 and-4 are 57% identical whereas BMP-3 is only 16-21% identical to BMP-2,-4,—5,-6 and-7 in the propeptide region. BMPs are a family of bone matrix polypeptides isolated from a variety of mammalian species. These proteins contain a highly conserved 7 cysteine transforming growth factor—beta domain in their C termini. Physical characteristics of the mature polypeptide subunits of BMP are listed in below table.

Unlike other BMPs, BMP-1 is not like transforming growth factor beta (TGF-β), but it is the prototype of a family of putative proteases implicated in pattern formation during development in diverse organisms. Although some members of this group, such as *Drosophila talloid* (TLD), are postulated to activate TGF-beta like proteins, but actual substances are unknown. Procollagenes C-proteinase (PCP) cleaves COOH- propeptides of procollagenes I, II and III to yield major fibrous components of vertebrate extracellular matrix. BMP-1 and PCP are identical.

Physical characteristics of mature subunits of BMP

Protein	No of Amino Acid	N-Glycos Residues	Mol. Wt (kDa)	Conserved cysteins	P.I
BMP-2	114	1	12.9	7	7.9
BMP-3	127	1	14.5	7	8.5
BMP-4	116	2	13.1	7	7.7
BMP-5	138	3	15.6	7	8.3
BMP-6	139	3	15.7	7	8.6
BMP-7	139	3	15.7	7	8.1

P.I—Isoelectric point

N-Glycos-Number of potential asparagine-linked glycosylation

Cellular events of bone formation: Sequence of cellular events that is induced locally by these extractable protein components of bone matrix has been characterized extensively and has been observed to be similar to that seen during fracture repair and during formation of embryonic endochondral bone.

During this sequence of events (i) mesenchymal cells are seen to migrate into BMP implanted site, proliferate after several days, and condense in regions, (ii) chondroblasts, believed to be derived from the early-appearing mesenchymal cells, form a cartilaginous template in the area of presumptive bone, (iii) at 10-14 days, cartilage hypertrophies, and cartilage extracellular matrix is vascularized by haematopoietic and endothelial cells, (iv) cartilage is gradually removed and replaced by bone, and at the end of 21 days an ossicle of bone, complete with marrow has been formed. This response is localized to the implant itself. However, the precise molecular and cellular mechanisms underlying bone induction remain to be elucidated.

Clinical application: Clinical utility of bone morphogenetic protein (BMP) has been explored by several investigators in animal systems and is presently under investigation in human trials. Clinical investigations of the use of demineralized bone matrix are performed in the hand, where demineralized allogenic bone implants are used to treat phalangeal cysts and in maxillo-craniofacial regions, where such implants are used to treat maxillo-cranial deformities and defects of the jaw. These early clinical reports are followed by controlled studies in which purified BMP is used to treat well defined critically sized defects in the long bones of animals.

Takagi and Urist (1982) used BMP alone and in combination of autogenous bone marrow to repair massive femoral diaphyseal defects in rats and observed union occurred more rapidly from transplant-implant composite than from autologous cortical bone. These observations suggested that bone marrow stroma cells include both mesenchymal cells and osteogenic precursor cells that respond to BMP by differentiating into osteoblasts for the repair of large femoral bone defects. Implants of BMP plus bone—marrow induced complete calvarial defect in dogs, suggested that calvarial bone regeneration is bone-marrow stroma-dependent for a supply of target cells. The target for BMP in cranial bone regeneration is the perivascular connective tissue cells (pericytes) of the host bone marrow stroma and endosteum. The molecular mechanism of differentiation of pericytes into osteoprogenitor cells is not known, but the process is irreversible.

Response to proto differentiated and differentiated bone cells to bovine bone morphogenetic protein (bBMP) is observed in implants in rabbit distal femoral metaphysis. Ca radioisotope quantitative methods, roentgenograms and histological sections shows that, by the end of fourth week, larger deposits of intra-metaphyseal cartilage and bone in bBMP than in control implanted (bovine serum albumin) femur and, by the end of eighth week, bone formation is associated with remodelling of the entire distal femur and expansion of the external diameter of the metaphysis.

It is observed that BMP/TCP (beta tricalcium phosphate) composite implant induces 91-100% incorporation by deposits of new bone in comparison to control implants of TCP impregnated with bovine serum albumin (BSA/TCP) induced 0-8%, in the skull trephine defects of a critical size of 1.4 cm in adult dogs that would otherwise remain unhealed in the lifetime of the individual. In another study, Ferguson et al (1987), who have created bilateral cranial trephine defects (14-20mm) in Rhesus monkeys and found that BMP implanted defects, produced more complete regeneration than the control implants (bovine serum albumin). The tissue responses are analyzed by computed histomorphometric and routine histological examination. The morphogenetic response occurred in the following sequences: mesenchymal cell proliferation, chondrogenesis and increased bone formation. Similar observations are also found where bBMP implanted in 18-20 mm trephine skull defects in adult sheep. In two out of five sheep, trephines are repaired with bone as early as 4 weeks postoperatively and in other three sheep, repair is complete after 8-12 weeks postsurgery.

Molecular cloning of BMP and their subsequent expression in recombinant systems has permitted the use of their osteoinductive activities in a variety of experimental model. Implantation of the recombinant protein in rats shows that a single BMP can induce bone formation *in vivo.* A dose response and time-course using the rat ectopic bone formation assay reveals that implantation of 0.5 - 115µg of partially purified recombinant human BMP-2A (rhBMP) resulted in cartilage by day 7 and bone formation by day 14. The time at which bone formation occurred is dependant on the amount of BMP-2 implanted; at high dose bone formation could be observed at 5 days. The cartilage and bone inductive activity of rhBMP is histologically indistinguishable from that of bone extracts. In this study, rhBMP-2 is combined with carrier matrix and implanted into a subcutaneous site in rats. It induced complete sequence of endrochondral ossification, including local formation of cartilage that is removed and replaced with bone in the period of 14 days. Implantation of a very large amount of rhBMP resulted in concurrent formation of cartilage and bone. These findings suggest that the molecule may influence intramembraneous pathway as well as endochondral sequences of bone formation. They also reports that most of BMPs (BMP-2 to BMP-7) induced bone in a very similar fashion and the specific activities of different molecules resulted in different amount of cartilage and bone formation.

Subcutaneous implants of recombinant human bone morphogenetic protein (rh-BMP-2) in rats have resulted in local induction of endochondral bone formation. To test the osteoinductive activity of rhBMP-2, they create five millimeter segmental defects in femora of rats. Using two doses (1.4 and 11.0µg) of lyophilized rhBMP-2 in combination with an inactive demineralized rat bone-matrix carrier, they reveals that both doses of rhBMP-2 induced formation of endochondral bone in the osseous defects in a dose-related manner. Implantation of high doses (11.0µg) of rhBMP-2 yields significant bone formation, resulting in radiographic, histological and mechanical evidences of union. Despite new bone formation in the defects that had received smaller doses (1.4µg) of rhBMP-2, no evidence of bone union is observed. In another experiment, rhBMP-2 is used in a 2.5 centimeter osteo periosteal segmental defect in the femora of sheep and stabilized with bone plate. The animals are treated with rhBMP-2 and a carrier, the carrier alone, or an autogenous graft, or they are not treated. Radiographic evaluation shows osseous union of all defects that have been treated with rhBMP-2 and of bone defects that have been treated with an autogenous bone graft. No union is observed in other groups. Biochemical testing shows that one month after implantation, the femora that have been treated with rhBMP-2 or an autogenous bone grafts have bending strength that is comparable with those of contra lateral femora.

Osteoinductive activity of other forms of BMP has also been studied. In one experiment, the osteoinductive activity of recombinant human osteogenic protein-1 (rhOP-1=BMP7) compare with that of naturally occurring bovine osteogenic protein, inactive collagen carrier and no implant in 1.5 cm segmental defect of rabbit ulna. All implants induce complete radiographic osseous union within eight weeks. The bone induces by both types of implants have new cortics with advanced remodelling and marrow elements. Histological evaluation of this new bone at eight weeks postoperatively reveals primary lamellar bone, with the formation of new cortics and normal appearing marrow elements. The control defects that have been implanted with collagen carrier alone and those with no implant shows no bridging of the defect.

Bone morphogenetic protein-2 is a low molecular weight glycoprotein, classified as morphogens. BMP-2 has pleiotropic functions that range from extra skeletal and skeletal organogenesis to bone generation and regeneration. BMP-2 induces structurally sound orthotropic bone in a variety of experimental systems, including femoral defects in rats, tibial and ulnar defects in rabbits, femoral defects in sheep, mandibular defects in dogs, spinal fusion in dogs and porous ingrowths in rats. Although the action of BMP on osteoblast differentiation has been extensively investigated, its effect on osteoclasts differentiation remains unknown. Kanatani et al (1995) examined *in vitro* effects of BMP-2 on osteoclasts like cell formation and bone resorption. BMP-2 (1-100ng/ml) significantly simulates bone resorption by pre-existent osteoclasts-like cells in mouse bone cell cultures containing stromal cells, whereas it do not affect the bone resorbing activity of isolated rabbit osteoclast-like cells. When BMP-2 is added to unfractionated bone cells after degeneration of preexistent osteoclast like cells, BMP-2 dose dependently simulates osteoclast like formation at a minimal effective concentration of 10pg/ml. BMP-2 is also enhanced the osteoclast like cell formation induced by 1, 25-dihydroxy vitamin-D3. Their findings indicate that BMP-2 stimulates bone resorption through both direct stimulation of osteoclast formation and activation of mature osteoclasts, possibly via stromal cells, *in vitro.*

Osteogenic protein-1 (OP-1) referred to as BMP-7, implanted in conjunction with bone derived type 1 collagen or various non-proteinaeous biodegradable carriers into surgically created large diaphyseal segmental defects in animals, leads to regeneration of new bone that is fully functional biologically and biomechanically. Further study has shown that OP-1 can be used as a bone graft substitute to promote spinal fusion and to aid in incorporation of metal implants. Osteogenic protein-1 (OP-1/BMP-7) also has been shown to be expressed temporally and spatially during epithelial mesenchymal interaction mediating tissue morphogenesis in early embryogenesis. Their experiments indicate OP-1 mediates the inductive signal involved in mammalian eye development. This report also supported by several researchers. BMP-7/OP-1 is necessary not only for mammalian eye developments but also for nephrogenesis and skeletal patterning. BMP-7 deficient mice died shortly after birth because of poor kidney development. Histological analysis of mutant embryos at several stages of development reveals that metanephric mesenchymal cells fail to differentiate, resulting in a virtual absence of glomerulus in newborn kidneys. This identified BMP-7 as an inducer of nephrogenesis. In addition, BMP-7 deficient mice had eye defects that appear to originate during lens induction and also have skeletal patterning defects restricted to the rib cage, the skull and the hind limbs. Overall, BMP-7 is an essential signaling molecule during mammalian kidney and eye development. Ashina et al (1996) reported that osteogenic protein-1 (OP-1/BMP-7) induces chondroblastic, osteoblastic and adipocytes differentiation to clonal murine target cells. OP-1 can induce differentiation along elements of the endochondral ossification pathway according to the stage and potential of the target cell. OP-1 induces endochondral bone formation *in vivo*, but their response is dose-dependent. Dose response studies with rhBMP-2 suggest that although the molecule by itself is capable of inducing bone formation *in vivo*, 10-fold more rhBMP-2 is necessary to achieve the same level of bone inductive activity observed with neutral bBMP.

Embryonic induction, the process by which signals from one cell population influence the fate of another, plays an essential role in the development of all organisms. In many cases, the signaling molecules belong to TGF-beta super family. Genetic studies in Drosophila on TGF-beta related gene, decapentaplegic (dpp), reveals the existence of conserved mechanisms regulating both the expression of the protein during developmentantal way in which interacts with other signaling molecules to generate pattern within embryonic tissues. BMP-4 (also known as (2β) is closely related to Drosophila decapentaplegic (dpp) gene, and like BMP-4, dpp is excluded from the neurogenic region. BMP-4 functioned as an analogous role to that of dpp in Drosophila, suggesting that the mechanism of dorsal-ventral patterning in xenopus and Drosophila embryos may be conserved. BMP-4 stimulates erythroid differentiation in ventral mesoderm during xenopus embryonic development. BMP-4 is known to be a key regulator of ventral mesoderm as well as specification of ectoderm development in xenopus embryo. BMP-4 gene had been isolated from a mouse genomic library and from a human placental cDNA library. Highly purified preparation of osteogenin (BMP-3) has been demonstrated to induce bone formation in an *in vivo* system.

To develop a controlled release system for bone morphogenetic protein (BMP), encapsulation of BMP with PLGA (Poly-DL-lactide-Co-glycolide) and PEG (poly ethylene glycol) are examined by Isobe (1995). PLGA capsules released much BMP compared to PEG capsules. These capsules containing BMP are implanted in subcutaneous areas of rats to examine the osteoinductive activity *in vivo*. They are digested completely, however, only PLGA capsules could induce new bone formation, including bone marrow, three weeks after implantation. It is then considered that PLGA capsule containing BMP is a prominent system to induce bone in clinics. Identification of hBMP-6 in placenta and brain cDNA libraries, OP-1(hBMP-7) in placenta and several other cDNA libraries, and multi tissue expression patterns of murine Vgr-1and murine BMP-2 indicates that BMP family of proteins are likely to have other growth and differentiation activities unrelated to bone induction.

[For full references: Maiti S K and Singh GR (1998). Bone morphogenetic proteins (BMP) - Novel regulators of bone formation: A perspective Indian J Expt Biol 36: 237-244].

8

Transforming Growth Factor Beta (TGF-β) in Bone Remodeling

Bone differs from other tissues not only in physio-chemical structure, but also in its extraordinary capacity for growth, continuous internal remodelling and regeneration throughout post fetal life. Bone formation is a complex process regulated by diet, vitamins, hormones and growth factors. Growth factors are polypeptides that increase cell replication and have important effects on differentiated cell function. Growth factors are initially considered as systemic agents, but current evidence indicates that they act primarily as local regulators of cell growth. The transforming growth factor-β family of proteins, which comprises polypeptide growth factors that have diverse effects on the growth, differentiation, and function of cells, is receiving considerable attention for potential clinical applications and is likely of physiological and orthopedically significance.

Transforming growth factors

It is one of the most important growth-promoting osteo-inductive substances, which have been identified at the site of fractures and other places. Transforming growth factors (TGFs) have been categorized into TGF-α (alpha) and TGF-β (beta). TGF-α has not been isolated from bone tissue and cannot be considered a local regulator of bone remodelling, although it is mitogenic for bone cells and stimulates bone resorption. TGF-α shared amino acid sequence, receptor-binding and functional activity with epidermal growth factors but had only weak transforming growth factors activity. Furthermore, TGF-β by itself is completely ineffective, but together with TGF-α, it potently induced growth of colonies of cell-line.

TGF-β is a multifunctional growth factor that has been shown to mediate normal cellular physiology and tissue embryogenesis and to participate in a variety of responses associated with inflammation and tissue repair.

Transforming growth factor-beta

Sources

TGF-β is originally purified from human platelets, human placenta and bovine kidney. The largest source of TGF-β in the body is the extracellular matrix of bone and platelets may represent the second largest reservoir for the peptide.

Chemistry and types

TGF-β is a highly stable molecule that consists of two identical chains, each containing 112 amino acids. TGF-β is a family of polypeptide growth factors encoded by closely related genes. There are at least five TGF-βs: TGF-β_1 to TGF-β_5. TGF-β_1, β_2, and β_3 have been found in many species, including humans; β4 has been found in chickens and β_5 has been found in amphibians. All share 64-82% similarity in their amino acid sequence. Human platelets contain a single form of TGF-β composed of two identical polypeptides constituting a dimer of relative molecular mass (Mr) 25,000.

A very high level (more than 95%) of sequence homology for a single isoform of transforming growth factor-β among many species indicates that these five isoform are not simply species-specific variants. All the five isoform might be expressed within a single species, in which all or a subset of the transforming growth factors βs may be synthesized by particular tissue at specific stages of development, or, after appropriate stimulation. A number of biochemical featured is shared by most of these closely-related proteins. The mature TGF-β subunit contains two identical chains, each containing 112 amino-acids (except TGF-β_4) 9 cysteine residues whose locations are conserved in all five isoform. The active proteins are approximately 25 kDa and are composed of two, usually identical, disulphide-linked subunits. Physical characteristics of the mature polypeptide subunits of TGF-βs are listed in the below table.

Physical characteristics of the mature polypeptide subunits of TGF—βs

Protein	Amino acids per Subunits	Molecular weight (Mr) (Da) cystine	Conserved
TGF-β_1	112	12,500	9
TGF-β_2	112	12,500	9
TGF-β_3	112	12,500	9
TGF-β_4	114	12,900	9
TGF-β_5	112	12,500	9

TGF-β is synthesized in precursor form that requires proteolytic processing, It is released from the cell in an inactive high molecular weight complex composed of the processed dimer in combination with amino terminal fragments of each of the two TGF- β monomeric precursors and a third uncharacterized molecule. The physiological mechanism by which TGF-β is released from the inactive complex is uncertain but may require specific enzyme activation.

Cellular activity

Transforming growth factor-β has a broad range of cellular activities, including the control of the proliferation and expression of the differentiated phenotype of several types of cells specific to the skeleton, among them the pluripotent undifferentiated mesenchymal tells (pericytes) for chondrocytes, osteoblasts and osteoclasts. *In vivo* and *in vitro* studies have demonstrated the synthesis of TGF-β by chondrocytes and osteoblasts and their expression is increased in fracture callus and specifically regulates bone-associated activities that may be important for the initiation of repair of fracture. Studies on the potential role of transforming growth factor β in soft tissue wound healing have shown that this molecule is released from degranulating platelets at the site of an injury, possibly to initiate a cascade of reparative events. Histologically, TGF-β increases nuclear 34-thymidine labeling in the osteoblast precursor cell zone. Studies with isolated bone cells show that TGF-β is a potent regulator of osteoblastic cell activity and on a molar basis TGF-β is one of the most effective mitogens so far described for osteoblast-enriched cultures from fetal bone.

Studies have also shown that TGF-β is the most potent chemotactic factor released from macrophages. In addition to attracting macrophages to sites of injury, this peptide activates the synthesis of other growth factors and deactivates the production of hydrogen peroxide, by macrophages, so that young tissues are not destroyed. Moreover, TGF-β has been shown to stimulate fibroblast-directed tissue repair by upregulating the production of extracellular matrix components, such as collagen, fibronectin and proteoglycans, by upregulating the expression of cellular integrin receptors for extracellular matrix proteins; and by inhibiting the action of proteolytic enzymes that could destroy newly-formed osseous tissue.

TGF-β produces pluripotent and biphasic changes in the biochemical activities of cells at particular stages of differentiation within the osteoblast lineage. To date most studies to evaluate the effectiveness of TGF-βs in bone are performed with prototypical isoform of TGF-β derived from blood platelets, but TGF-β_1, β_2, β_3 each have identical qualitative effect in osteoblast enriched cell cultures.

All the three isoforms potently increase synthesis of bone DNA at low concentration but have reduced mitogenic activity at higher levels. These isoforms also enhance synthesis of collagen and non-collagen protein and decrease activity of alkaline phosphate in osteoblast enriched cultures, but *TGF*-β_3 appears to be 3-10 times more potent than TGF-β_1 and β_2. Similarly, the TGF-βs enhance replication of cells and production of bone matrix in cultures of intact bone *in vitro.* In most instances, the TGFβs enhance synthesis of type-l collagen and non-collagen protein in bone cells. The mitotic response to TGF-β is biphasic; low concentrations (below 100rM) are stimulatory, whereas higher levels produce less of a stimulatory effect. At higher, less mitogenic concentrations, TGF-β alters expression of various activities associated with the osteoblast phenotype in different ways; under these conditions, TGF-β decreases alkaline phosphatase activity and similar to its effects in a number of other connective tissue systems, enhances the synthesis of type I collagen, the major organic element m the bone matrix. Furthermore, TGF-β_1 and β_2 enhance chemotaxis of osteoblasts, suggesting an additional role involving the recruitment of differentiated osteoblasts that may be necessary for new bone formation and fracture repair.

Recent studies in osteoblast-enriched cultures have demonstrated that TGF-β_1 increases synthesis of type I collagen polypeptide in the absence of mRNA transcription and decreases non-collagen in the bone matrix as a result of the action of TGF-β probably occurs by multiple transcriptional, post—transcriptional and post-translational events. Some other activities of TGF-βs associated with the function of osteoblasts have also been examined. TGF-β_1 and β_2 increase the transcriptional and polypeptide levels of osteopontin, a protein in the bone matrix that is thought to be important in adhesiveness of bone cells. In osteosarcoma cultures, TGF-β enhances Type 1 collagen synthesis as well as the production of a variety of bone matrix-associated polypeptides such as osteonectin, osteopontin and alkaline phosphatase, but decreases osteocalcin synthesis. These findings indicate that TGF-β enhanced bone matrix accumulation and consequently the maintenance of bone mass. In general, the stimulatory influence of TGF-β on bone matrix protein synthesis may be attributed to effect at the transcriptional level.

However, a direct relationship between the influence of TGF-β on type 1 collagen mRNA and polypeptide levels is not observed in isolated bone cells, suggesting that TGF-β has an additional supportive rote in increasing bone matrix formation. The protein in the bone matrix termed osteonectin, which may be involved in the deposition of type-1 collagen and the transition between cartilage and bone, is increased by TGF-β_1 in an in *vitro* model that is used to study repair of fracture.

Much less information is presently available regarding direct effects of the TGF-βs on osteoclasts or osteoclast precursors. Recently, biphasic effects on the development of osteoclast- like cells have been reported, in which low concentrations are stimulatory and higher concentrations are inhibitory. TGF-β_1 induces resorp-tion of bone on synthesis of prostaglandin E_2.

In fetal rat long bones, TGF-β decreases bone resorption, and this effect may be related to the ability of TGF-β to inhibit the formation of osteoclast-like cells *in vitro.* This difference may be a significant distinction in bone growth and development between the fetus and the new born; within very early stages of bone formation, matrix formation may be more important to the organism than bone remodelling. It is therefore suggested that TGF-β has an additional, indirect role in increasing or maintaining body mass.

With regard to formation of cartilage, the TCF-βs ap-pear to increase differentiation of mesenchymal cells, production of proteoglycans, and replication of chondroblasts. In contrast, these agents usually decrease the proliferation and function of differentiated cells by more mature chondrocytes. For example, TGF-βs decrease the activity of alkaline phosphatase. The highest levels of TGF-β mRNA are associated with osteoblasts in developing bone, whereas TGF-β transcripts decreased in chondroblasts /chondrocytes with increased type II collagen expression.

In general, it seems that one important role for the TGF-βs is to include developmental transitions in cells that are involved in formation of endochondral bone at particular stages of differentiation. The predominant trend of these effects appears to be toward eventual *de novo* formation of bone.

Clinical applications

The use of *TGF*-βs to enhance wound healing is first investigated by Mustoe et al (1987), who applied this peptide directly to and immediately after the creation of linear incisions through the dorsal skin of rats. These investigators demonstrate a 220% increase in the maximum strength at the site of the wound after five days and acceleration in the rate of healing by at least three days compared with that in the control animals.

It is also suggested that this peptide molecule has the capacity to stimulate both intramembraneous and endochondral formation of bone. Noda and Camilliere (1989) studied the effect of daily injection of 1 μg of TGI-β directly into the periosteum of the parietal bones of neo-natal rats. They reveal that (i) the thickness of the treated parietal bones increases approximately twofold; (ii) this effect is localized to the site of the injection. (iii) no changes are observed in the contra lateral bones or at distant skeletal sites.

Joyce *et al* (1990) performed a similar study in the femora of newly-born rats and found that on daily injection of either TGF-β_1 or β_2 into the subperiosteal region of the femora, mesenchymal -precursor cells in the periosteum are stimulated to proliferate and differentiate much the same as is observed during the embryological formation of bone and early fracture healing. After cessation of the injection, endochondral ossification also occurred and this resulted in the replacement of cartilage with bone In addition, it is shown that injection of TGF-β_2 stimulated the synthesis of TGF-β_1 in chondrocytes and osteoblasts, suggesting auto regulation of this peptide.

It has been demonstrated that a single injection of 25-100 nanogram (ng) of recombinant human TGF-β_1 adjacent to the ear cartilage of rabbits stimulated the formation of bone alter 21 days. In a subsequent investigation, researcher's shows that exogenously applied recombinant human TGF-β stimulated the recruitment and proliferation of osteoblasts in critically-sized defects in the skull of rabbits. By assessing the temporal dynamics of the formation of bone in these defects, they demonstrate the potent osteoinductive activity of this peptide and its potential therapeutic applications for non-healing osseous defects.

The function of TGF-β, both *in vitro* and *in vivo* can be highly variable. In some instances this growth factor inhibits cell growth and matrix synthesis, whereas in others it stimulates these processes. These different functions depend on the target cell, the presence or ab-sence of other cytokines and the dose. In their study, Sumner et al (1995) suggest that the response to treatment with TGF-β is dose dependent, with the lower dose being in more effective for enhancing bone ingrowths, whereas, high dose of TGF-β inhibit mineralization. Synthesis of osteocalcin, a calcium binding protein important in mineralization is inhibited, possibly due in this growth factor.

A similar observation is also reported by Nielsen et al (1994) *who* injected TGF-β, 4 to 40 ng in every alternate day for 40 days after creation of tibia fractures in rats. Mechanical testing shows that TGF-β induced a dose-dependent increase of the callus at the site of the fracture.

Enhancement of bone ingrowths with TGF-β is also evaluated in the canine model. Sumner et al (1995) observed that treatment of an implant (titanium fiber metal-coated rod with a combination of a hydroxyapatite-tricalcium phosphate coating and TGF-β may result in better bone ingrowths than that obtainable from grafting of the gap with autogenous cancellous bone, The amount of new bone formation in three millimeter gaps adjacent to the treated implants is twice that in the gaps of the paired controls, regardless of the dose. The differences between the treated and control implants with regard to architecture of the new bone in the gap indicate that the mechanism of action of TGF-β_1 may include both proliferation of osteoprogenitor cells and production of matrix by committed osteoblasts.

Histological study with two *in vivo* models of os-teogenesis have also demonstrated an increase in net formation of bone by TGF-β_1, and β2, when the agents are tested either by direct application to the tissue or by subcutaneous injection.

Summary

The transforming growth factor βs are polypeptide growth factory encoded by a family of closely-related genes that are expressed in numerous tissues and species. Bone is one of the first tissues in which locally-produced molecules with TGF-β like activity appeared to regulate normal cellular function, and the skeletal matrix probably comprises the largest reservoir of TGF-βs *In vitro* and *in vivo* studies have indicated that TGF-βs can have either stimulatory or inhibitory effects on replication, lineage development, and differentiated phenotypic function in many types of skeletal tissue cells. The effect may be biphasic in the sense that low concentrations of TGF-βs stimulate proliferations of cells, whereas high concentrations inhibit proliferation. Even within the same cell lineage, opposite effects have been noted with different concentrations of TGF-βs with skeletal cells at different development stages. In fetal tissue, TGF-β stimulates differentiation of mesenchymal cells, proliferation of osteoblasts and synthesis of bone matrix, but it may also induce maturation into and through the osteoblast, chondrocytes and osteoclast lineages. Most evidence to date leads to the conclusion that the TGF-βs have an important role in the formation of bone and this has tremendous potential in therapeutic application for non-healing osseous defects and in different types of fracture healing.

Finally, continued *in vivo* and *in vitro* studies of these molecules are needed in order to evaluate their effective doses, appropriate duration of treatment, tissue specificity, receptor-binding, intracellular signaling and the regulation of their activities by other local and systemic factors.

[For full references: Maiti SK and Singh GR (1996).Transforming growth factor-beta (TGF-β) in bone remodeling. Current Science, 71: 613-617].

9

In vitro Osteoinduction Potential of Silica Coated Hydroxyapatite Bioscaffold Seeded with Mesenchymal Stem Cell

The knowledge and principles from interconnected disciplines of bioengineering, material science and life sciences are combined in tissue engineering with the aim of developing a construct, that can wholly or partly restore/maintain/augment the function of a damaged tissue or organ. Tissue engineering construct generally comprises cells on a suitable scaffold. The ultimate functionality of a tissue engineering construct is mainly dependent on the cell behavior on scaffolds. The proliferation and cell viability on scaffolds varies from one cell type to another and from one species to another. So it is important to investigate individual cell-scaffold combinations. The similarity in structure and composition to bone mineral, osteoconducive properties, ability to integrate with the bony tissue and absence of immune response makes hydroxyapatite $[Ca_{10}(PO_4)_6(OH)_2]$ the choice of calcium phosphate biomaterial with only disadvantage of slow rate of resorption. On contrary, silica–calcium phosphate composite in comparison to calcium phosphate is rich biomaterials having a faster resorption rate. Studies have shown that coating of hydroxyapatite with a calcium silicate containing layer encourages cell proliferation and osteogenic differentiation of human bone marrow-derived stromal cells. So in this chapter, a triphasic composite scaffold (calcium silicate, hydroxyapatite and tricalcium phosphate), namely HASi with elements in the following percentages: 66.36 % - calcium, 25.35 % - phosphorus and silicon - 8.29 % and porosity of 50 – 500 μm is reported to evaluate the rBMSC proliferation and differentiation properties *in vitro*.

The present chapter is therefore, included with the aim of assessing cytocompatibility of HASi and rBMSC in terms of cell attachment, cell morphology, cell proliferation and osteogenic differentiation.

Bone marrow aspiration, isolation and culture of rabbit bone marrow derived mesenchymal stem cells (rBMSC) are done as per previous published reports (Maiti et al., (2012). For more detail: "Stem Cell Research in Lab Animals"

(2020) by Swapan Kumar Maiti. Third passage of rBMSC is used for this study.

Quantification of mesenchymal progenitors (CFU-F Assay)

For CFU-F assay, 1×10^6 bone marrow nucleated cells per 60-mm dish are seeded and cultured at 37 °C in 5% CO_2 for 12 days with media change on every 2 days, followed by crystal violet staining. The numbers of CFU-Fs are counted under a microscope.

Quantification of osteoblasts (CFU-OB Assay)

For CFU- OB assay, 1×10^6 bone marrow nucleated cells per 60-mm dish are seeded and cultured for 21 days in differentiation medium (α-MEM containing 10% FBS, 1% penicillin/streptomycin, 50 μg/ml L-ascorbic acid and 2.0 mM b-glycerophosphate). Media is changed on every 2 days. Alizarin red staining is done to identify the mineralized bone matrix of osteoblast differentiated colonies, designated as CFU-OB colonies.

Osteogenic differentiation

Undifferentiated cells are induced towards the osteogenic lineage. Putative MSC is seeded in six well plates at a density of approximately 3000 cells/cm^2 and cultured in expansion media till it reached 90-100% confluence. Osteogenic differentiation is induced by culturing the cells for 20 days in osteogenic induction medium (OS) consisting of 100nM dexamethasone, 10mM β-glycerophosphate, and 0.05 mM L-ascorbic acid-2-phosphate and 10% FBS in low glucose DMEM. As a negative control an equal number of wells are maintained in expansion media for 20 days. The media in both groups is replaced completely every 4 days. The cells are cultured at 37^{0}C in a CO_2 incubator at 5% CO_2 and maximum humidified environment. The presence/differentiation of osteoblasts is confirmed by mineralization staining (Alizarin red S staining) and alkaline phosphatase (ALP) activity.

Fluorescent staining of rBMSC-scaffold constructs

For assessment of cell viability and attachment on scaffold, the cell–scaffold constructs are rinsed twice with sterile PBS and submersed in 3mL of PBS. CFDA-SE is then added to the PBS and incubated for 15 minutes at 37^{0}C to label live cells green (emission at 517 nm) by labeling intracellular esterase as recommended by the manufacturer. They are rinsed twice again with PBS. Propidium iodide (PI) (20 μg/ml, Molecular Probes) in PBS is added for 2 min at room temperature followed by washing with PBS. All steps are performed in the dark and samples are then viewed under a fluorescence microscope.

Scanning electron microscopy (SEM) study

Attachment properties and morphology of rBMSCs from third passage are assessed on HASi blocks of 0.3x0.2x0.1cm dimension. For this, the scaffold blocks are coated with fibronectin for 24 h and then surface seeding of third passage rBMSCs is done with 1x 10^6 cells per mL of DMEM medium containing 10% fetal bovine serum and antibiotics (penicillin G, 100U/ml; streptomycin, 0.1 mg/ml; amphotericin B, 0.25 μ g/ml) in six well plates followed by incubation at 37°C in a humidified 5% CO_2 environment. Unseeded scaffolds are also maintained in the same culture conditions as control. The cell-loaded scaffolds and control are rinsed with PBS after 3 and 17 days of cell seeding respectively, and fixed with glutaraldehyde 2.5% for 1 h. Then the scaffolds are dehydrated with a graded ethanol series (30 %, 50 %, 70 %, 90 % and 100 %) and incubated for 10 minutes in hexamethyl- disilazane (HMDS) and dried overnight in a desiccators, and are gold coated. They are observed under scanning electron microscope (Jeol, Japan).

Evaluation of calcium crystals

The ability of cells to produce mineralized matrix and calcium crystals is a key factor for bone regeneration. To determine the formation of calcified extracellular matrix deposition due to mineralization phenomena, a histochemical technique is used. Cells harvested from trypsinized scaffolds are fixed in 70% ethanol followed by washing with PBS and stained with 1 % alizarin red solution for 2 min. They are washed again with PBS and observed under a light microscope.

Histological evaluation

The decalcified scaffolds with Goodling Stewart's solution after day 17 post-seeding is maintained in the same culture conditions as for SEM analysis. They are subjected for histological study with H & E staining to evaluate the bone formation ability. Bone marrow collection from rabbits is carried out as per the standardized procedure. The mesenchymal stem cells are successfully isolated, cultured and propagated from all the aspirated bone marrow samples. Third passage of rBMSC is used for seeding in HASi scaffold.

From CFU-F assay, 52±2.3 (mean ± SD) colonies of fibroblast are obtained from three different trials and from CFU-OB assay, 49±2.6 (mean ± SD) colonies of osteoblasts are obtained from an initial seeding density of 1 X 10^6 cells from buffy coat and third passage respectively in 60 mm culture dish.

To characterize unique differentiating property of rBMSCs and mineral deposition, osteogenic differentiation is performed in monolayer culture for 3 weeks. The cells are changed from a fibroblastic appearance to a more cuboidal appearance at day 9 of culture. Proliferation and differentiation of control and OS cultures are compared using inverted microscope. In both cultures, cells are proliferated and reached almost complete confluence at day 9. In OS cultures, nodular aggregates of cells are evident at day 9 of culture and increased up to 21 days. These aggregates are characterized by deposits of amorphous material. In control cultures aggregates are not observed and deposits are also lacking. The OS cultures are stained positive (orange red) with alizarin red S stain for mineral (calcium) deposition in their newly formed matrix. Alizarin red positive nodular aggregates present at day 21 are larger and stained more intensively, indicating that more extensive calcium depositions have occurred. In the control wells, adherent monolayer of swirling spindle shaped cells is over confluent without any cell aggregates and is stained negative for mineral deposition. ALP activity or expression is higher in osteogenic differentiated cells than control culture. In positive cases the nucleus of MSCs taking bluish-purple colour.

Fluorescent staining of rBMSC-scaffold construct reveal that the live cells emitted green fluorescence. On day 3 post seeding, no dead cells are seen.

SEM observation of the seeded scaffold on day 3 post-seeding is shown stem cell attachment on the surface of HASi. They have a wavy surface and the attachment is close but not tight. On day 17 post-seeding, the attached stem cells show expansion and proliferation to form cell sheet-like structure with strong anchoring to the scaffold surface by footpads. There is matrix deposition in the form of crystals over and around the attached stem cells. The cells are seen closely attached to the side wall of the pores. The panoramic view of calcium matrix deposition by the attached cells is similar to snow covered caves. The adjacent crystals physically entangled to form a strong interconnected network that shows similarity with calcium orthophosphate. The control HASi maintained in same culture conditions for the same time period as 17 days post-seeding samples; do not show any crystal deposits.

Trypsinization and subsequent alizarin red staining of the crystals deposits from day 17 post-seeding sample of HASi shows intense red staining indicative of presence of calcium. The histological sections stained with H& E staining shows pink stained areas are indicative of beginning of early new bone formation.

Conclusion

Bioscaffold HASi has the potential to induce osteogenic differentiation of rBMSCs resulting in calcium matrix deposition in the absence of osteogenic induction medium. The rBMSC-HASi construct may be a promising bone substitute for autograft in non-healing and delayed fracture healing with potential bone loss. Further research is needed for understanding the cellular and molecular mechanisms of osteoinduction in this stem cell construct noticed in this study. Further research is also emphasized to investigate individual cell-scaffold combinations and investigate proliferation, viability and osteogenic differentiation of BMSC of different animal species on HASi scaffold for one ideal stem cell-based tissue engineering.

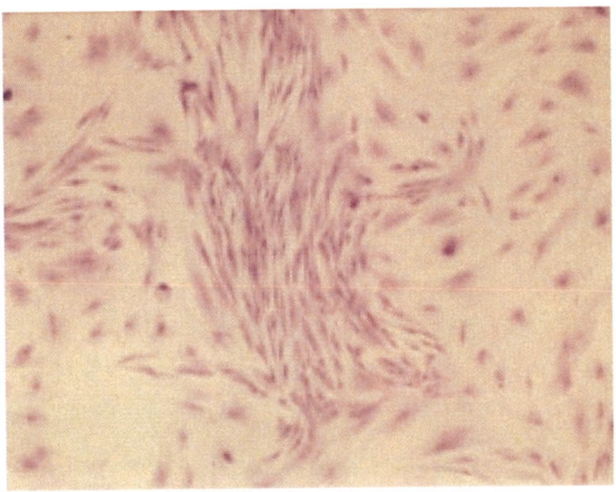

Fig. : Crystal violet staining of P3 rBMSC

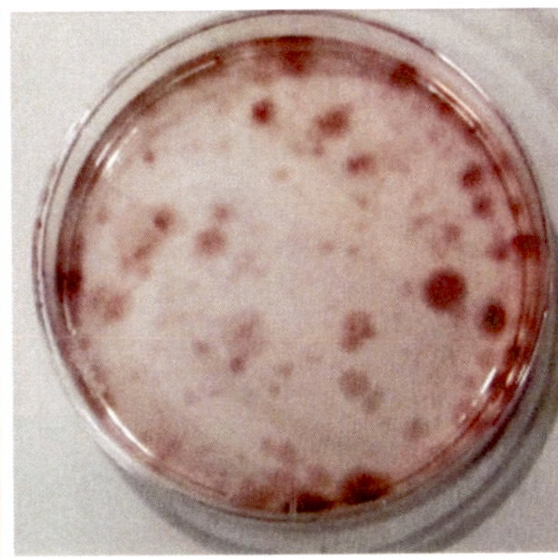

Fig. : Colony forming units of osteoblasts

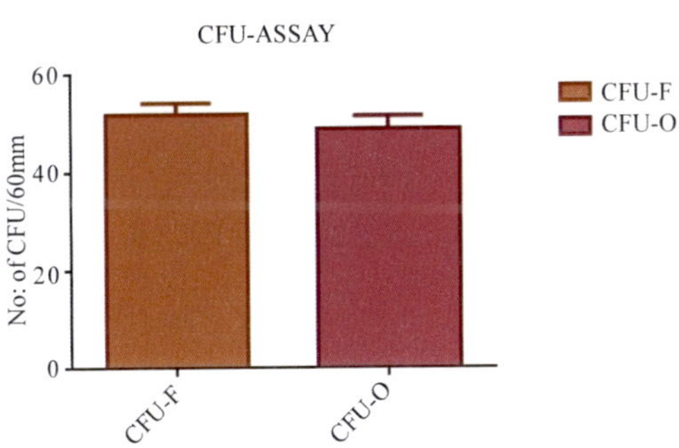

Fig. : Histogram showing number of CFU of fibroblasts and osteoblasts from an initial concentration of 1x 10^6 cells

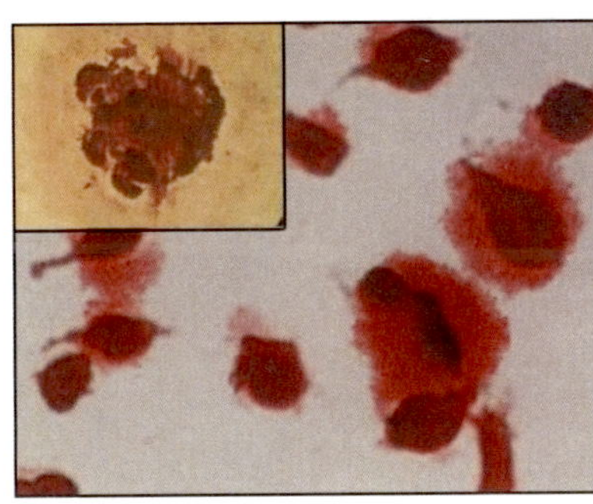

Fig. : Orange red staining calcium nodules representing osteoblastic differentiation of rBMSC

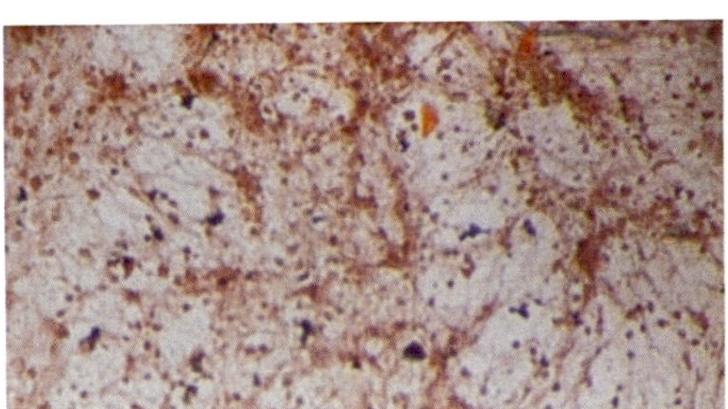

Fig. : Osteoblast with bluish purple nucleus after ALP staining

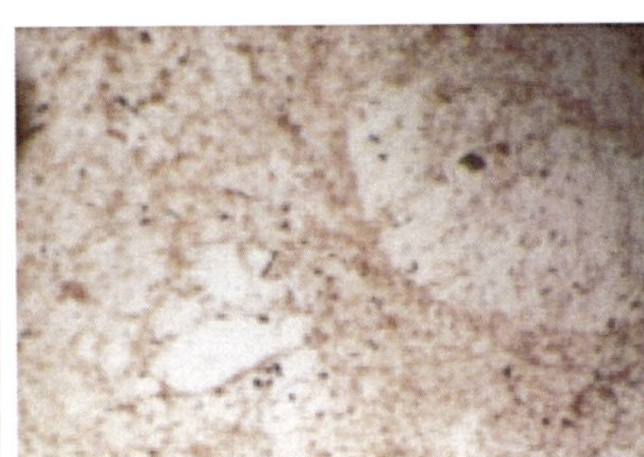

Fig. : Control well after ALP staining

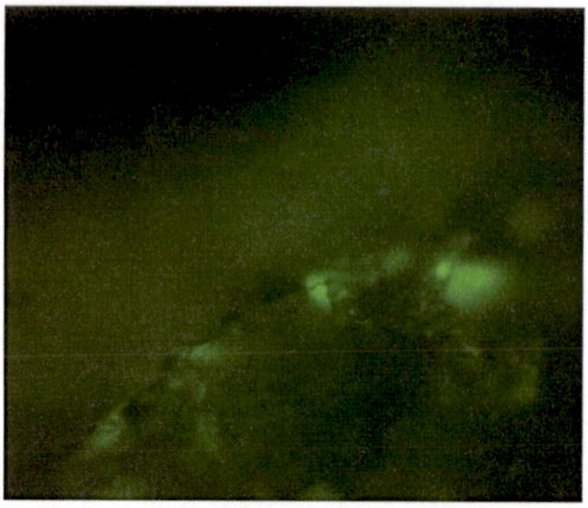

Fig. : Live cells emitting green fluorescence on day 3 post-seeding on the HASi scaffold

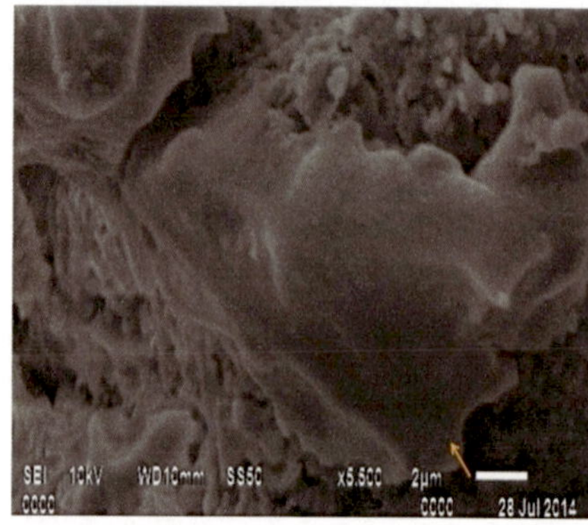

Fig. : SEM analysis of rBMSC seeded HASi on day 3 showing attachment of stem cell (arrow)

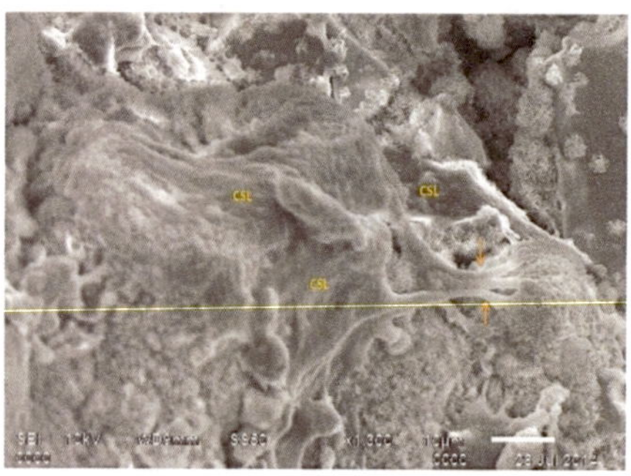

Fig. : SEM analysis of rBMSC seeded HASi on day 17 post-seeding showing cell sheet like structure (CSL) strongly anchored to scaffold surface by footpads (arrows)

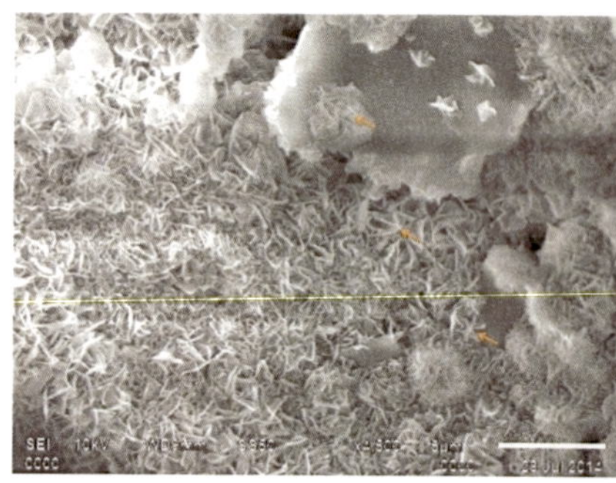

Fig. : SEM analysis of rBMSC seeded HASi on day 17 post-seeding showing matrix deposition (arrows) over and around the attached cells

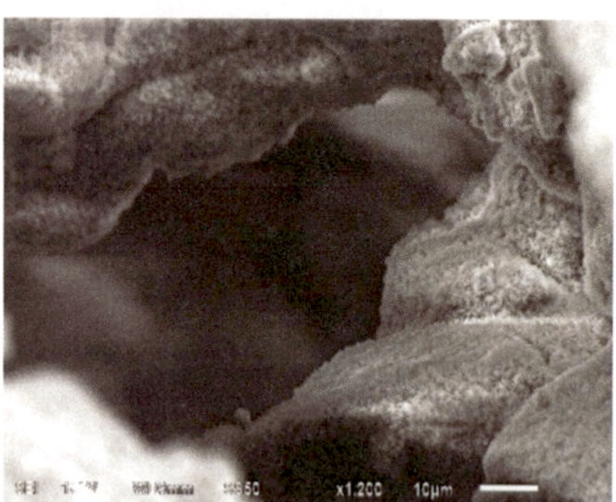

Fig. : SEM analysis of rBMSC seeded HASi on day 17 post-seeding showing a panoramic view of matrix deposition over attached cells

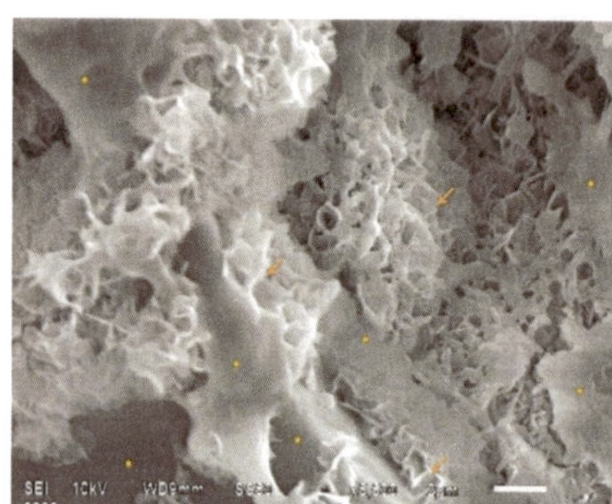

Fig. : SEM analysis of a closer view of rBMSC seeded on HASi on day 17 post-seeding showing physical entanglement of calcium orthophosphate like crystals (arrows) over and around attached cells (dots).

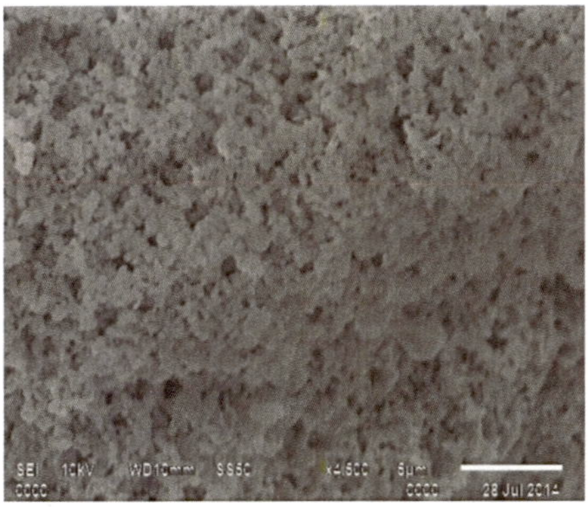

Fig. : SEM analysis of the unseeded HASi after 17 days

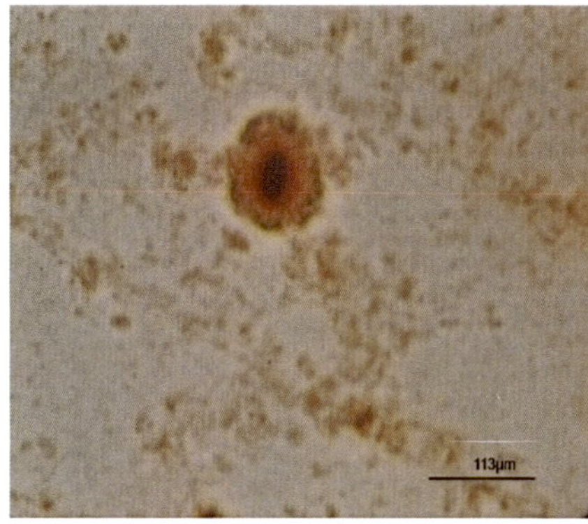

Fig. : Light micrograph of alizarin red stained calcium crystal trypsinized from day 17 post-seeding HASi showing positive intense red colour indicating presence of calcium

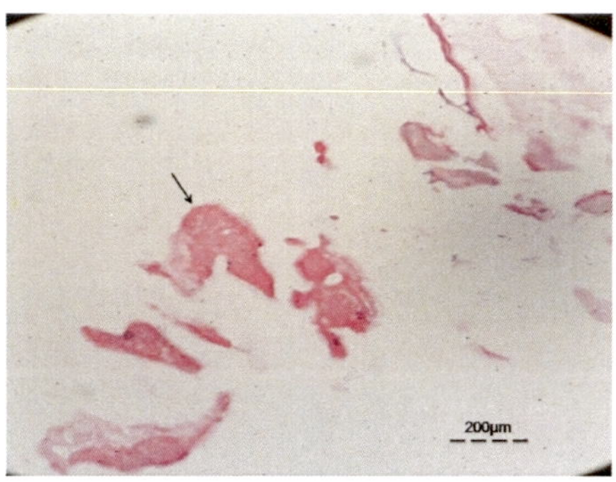

Fig. : H & E stained sections of decalcified day 17 post-seeded HASi scaffold showing dark pink stained areas (arrows) indicating early new bone formation.

10

Mesenchymal Stem Cells of Different Origin in Regeneration of Critical Sized Bone Defect

The process of fracture healing is unique, in which damaged bone restores its original architecture but it does not form a poorly organized replacement matrix, otherwise known as scar tissue, rather it regenerates the original matrix and retains its mechanical properties. In order to repair bone defects, a combination of cells with osteogenic activity and appropriate scaffolding material can stimulate bone regeneration and its repair. A potential substitute for autologous transplantation should possess 3 elements: osteoprogenitor cells, osteoinductive factors, and an osteoconductive scaffold. Stem cells from adult tissues are attractive materials for cell therapy and tissue engineering. These cells generally have restricted lineage potential when compared to embryonic stem cells, and this may be advantageous in certain therapeutic applications. Bone-marrow-derived mesenchymal stem cells (BMSCs) have proven to be beneficial in bone regeneration. Bone marrow stem cells are pluripotent cells with the capability of differentiating osteoblasts so that they have been used to facilitate bone repair. Several studies have shown that seeding of cultured BMSCs on bioabsorbable implants can induce bone formation *in vivo* and lead to improved healing of critical-size bone defects.

Mesenchymal stem cells (MSCs) because of their self replication and osteogenic differentiation capabilities are regarded as an excellent source of cells for bone tissue engineering. Autologous stem cells are not always preferable since the quality and quantity of such cells will be affected by metabolic diseases, old age, and osteoporosis. Allogenic BMSCs might be preferable to xenogenic, but they are always not readily available. In addition, allogenic MSCs have the potential for carrying some diseases. The use of non-autologous stem cells isolated from healthy donors offers a major advantage since these stem cells can be thoroughly tested and formulated into off-the-shelf medicine in advance. Major attractive advantage of BMSCs as a source of cell transplantation is their low immuno-genecity. It is now well established that BMSCs are immune-privileged cells that do not elicit immune responses due to an absence of their immunologically relevant cell surface markers.

In the last two decades, many of the bone substitute materials have been evaluated to replace the necessity for autologous or allogenic bone. Among them, the feasibility of bioactive ceramics, bioactive glasses, biological or synthetic polymers, and their composites for bone tissue engineering has been studied. Bioactive ceramics, namely calcium hydroxyapatite (HA), HA with Tri-calcium phosphate (HA/TCP) and bioactive glasses, have been used as scaffold for bone reconstruction for many years. They are termed 'bioactive' because they form an interfacial bond with tissues upon implantation and potentiate osteogenic activity as a result of surface modification when exposed to interstitial fluids. Bioceramic provides scaffold to support the attachment and migration of newly formed bone cells into the bone defect and also help in the formation of a capillary network through the newly formed bone. Bioceramic also acts as a carrier to deliver stem cells for osteogenesis and these bioactive inorganic materials are similar in composition to bone mineral that is of clinical importance.

Till now most of the xenogenic MSC transplantations for bone healing have been done from human to laboratory or large animal models of bone healing and between the two species of laboratory animals (rat and rabbit). The xenogenic transplantation of MSCs from small or large animal species to rabbit model of bone healing has not been attempted to the best of our knowledge. In this chapter, the potential of allogenic (rabbit) and xenogenic (sheep and dog) bone marrow-derived mesenchymal stem cells seeded on a biphasic bioceramic construct (namely HA/TCP) to enhance osteogenesis in a critical size bone defect (CSD) in rabbit model is presented.

The male adult skeletally mature New Zealand white rabbits of seven months to one year old and with a mean weight of 2.78 ± 0.36 kg are used in this study. They are maintained in separate cages, fed *ad libitum* and allowed to move freely during the whole period of study. For this purpose 36 animals are divided into six equal groups (each n=6). These groups are treated and compared as follows: the radial 15-mm critical-size bone defect is filled with HA/TCP seeded with autogenous rBMSC (group A); HA/TCP seeded with allogenic rBMSC (group B); HA/TCP seeded with ovine BMSCs (group C); HA/TCP seeded with canine BMSCs (group D), cell free HA/TCP scaffold only (group E). In group F, the critical-size bone defect is left as such (untreated) and served as control.

Hematological parameters are evaluated preoperatively in order to rule out systemic diseases. Based on clinical examination and preoperative radiographs it is assessed that the experimental animals have no signs of bone or joint disease on the relevant limbs.

A biphasic composite bioceramic comprising of calcium hydroxyapatite and tricalcium phosphate namely HA/TCP in the following percentage: calcium hydroxyapatite-65% and tricalcium phosphate 35%, and porosity of 200–450 µm is used as ceramic block. Each HA/TCP block is of 15 mm long and 5 mm in diameter are produced in an emulsion process.

Isolation, expansion and culture of rabbit BMSC (rBMSC), ovine BMSC (oBMSC) and canine BMSC (cBMSC) is performed as reported earlier (Maiti et al., (2012). For more detail: "Stem Cell Research in Lab Animals" (2020) by Swapan Kumar Maiti. The BMSCs of different origin are used at 3rd passage in this study.

The third passage cells (six million) of rBMSC, oBMSC and cBMSC are resuspended in 4 ml of Dulbecco's modified Eagle Medium (DMEM) and are transferred into a 5 ml tube. The HA/TCP bioscaffold are incubated at 4 ^{0}C in 50µM ml^{-}1 fibronectin solution diluted in phosphate buffered saline (PBS) for 24 h. The HA/TCP blocks are then placed into the medium containing the cells. After continuous spinning for 1.5 h at 37 ^{0}C (35rpm), the ceramic blocks are placed into a 6-well plate. The cell pellet obtained after centrifuging the media containing the cells at 800g for 5 min is resuspended in 70µl of culture medium and seeded on the ceramic blocks. This MSC seeded ceramic bioscaffold is then used in groups A, B, C and D. In group E, only cell free HA/TCP bioscaffold is used at the defect site.

All the animals are provided three weeks time period for acclimatization to the environment prior to surgery. Animals are given general anaesthesia with xylazine hydrochloride (6mg/kg, IM) and ketamine hydrochloride (60mg/kg, IM). Under strict surgical asepsis, a medial approach to the radius bone is performed and a unilaterally critical-size bone defects are created in the radial diaphysis. A 3 cm medial incision is given over the radius, soft tissue is dissected and the radius is exposed by gentle separation of the surrounding muscles. To protect the ulna, a Hohmann retractor is placed in between ulna and radius. A 15mm long osteo-periosteal critical-size bone defect is created with the help of an oscillating bone saw which is continuously irrigated with sterile cold saline to prevent thermal necrosis. The periosteum is removed from each side of the remaining radius. The bone defect is cleaned with sterile physiological saline and composite ceramic construct is placed into the bone defect as per treatment protocol for different groups. The stem cell- bioceramic construct is secured in position by suturing the bioceramic block in the bone defect with 2-0 absorbable sutures. The skin is sutured with non-absorbable nylon suture. A custom designed wooden splint padded with sterile gauze is applied on the test bone for one week postoperatively.

Regular dressing with antiseptic lotion is carried out for 5 days post surgery. Animals are administered ceftriaxone sodium (20mg/kg) intramuscularly for five days. Meloxicam (0.15 mg/kg, IM) is given for the first 3 postoperative days as an analgesic agent. The animals are kept in an individual cages and allowed unrestricted weight bearing activity post surgery. The animals are sacrificed after 90 days of experiment.

Radiographs are taken using table top procedure at 14mAs; 50kVp and 90 cm FFD. Craniocaudal and mediolateral radiographs of each bone defect site are taken immediately after surgery to monitor the position of the implants. Subsequently radiographs are taken at 14, 30, 60 and 90 days postoperatively to record the acceptance or rejection of stem cell-bioceramic construct and the initiation and progression of bone formation and healing at critical size bone defect site. The radiographs are graded with slight modification of the scoring system. Overall radiographic scores for each group are calculated by adding mean scores for each group at different time intervals. Resorption of bioceramic implant (HA/TCP) is qualitatively assessed.

The evaluation of new bone formation at defect site in different treatment groups is done by scanning electron microscopic (SEM) technique. The implants are explanted after euthanizing the animals on day 90 post surgery. Each specimen consists of both fracture healing site and adjacent bone. The samples are fixed in 2% glutaraldehyde phosphate buffer solution for 24-48 h. After fixing in glutaraldehyde, two washings for 30 min each with phosphate buffer saline (pH 7.4) followed by distilled water is performed. Then the specimens are dehydrated in serial concentration of ethanol 30-100% for 30 min each. Thereafter, the specimens are dried with Hexamethyldisilazine (HMDS) for 20 min. Specimens are mounted on aluminum stubs using adhesive silicon tape. After the golden ion sputtering is carried out on Jeol ion sputter Model JFC 1600 at 7-10mA and 1-2 kv for 15min. Finally, the specimens are observed under scanning electron microscope (Jeol JSM 6610 LV, Japan) at appropriate acceleration voltage and magnification range of the unit calibrated for the study of their different surfaces to identify the orientation and distribution of newly formed osseous tissue and distribution or absorption of construct at the bone defect site.

The test bones are collected from the animals sacrificed on day 90 postoperatively and a 2.5 cm long piece of radius including the site of the defect and normal host bone on both sides is cut using hexa saw. The bone segments are washed thoroughly with normal saline and fixed in 10% formalin for 48-72 h. The bone sections are decalcified in Goodling and Stewart's solution containing 15ml formic acid, 5ml formalin and 80ml distilled water. The sections are checked regularly for the status of decalcification. The completion of decalcification is assessed by flexibility, transparency and pin penetrability of the bone sections. The tissues are processed in a routine manner and 4μ thick longitudinal and transverse sections are cut and stained with Hematoxylin and Eosin (H&E) as per standard procedure. The stained sections are examined under different magnification in a light microscope for cellular reaction, healing process and fate of the implanted bioceramic. Bone healing is assessed in each group. Radiographic, histological and scanning electron microscopic examinations in this study are performed blindly.

In all experimental groups (A, B, C, D, E) except the control group (F), the food and water intake return to normal on day 3 while in the later on day 7. Weight bearing is mild to moderate on the operated limb from day 2 post surgery in all groups except group F. Complete weight bearing is seen on day 7 after surgery in all animals of groups A, B, C, D and E after removal of wooden splint. In the control group (F), animal does not bear weight normally during the study period. The implanted biomaterials (HA/TCP) do not shown any signs of inflammation, irritation or infection. The skin wounds heal uneventfully.

Due to overlapping of the radius by adjacent ulna bone in craniocaudal view, the radiographic evaluation of the defect site at different intervals mostly involve analysis of mediolateral views of the radiographs. The critical size defect is clearly visible with cut edges of the bone almost at the center of the radial diaphysis in all the animals of control group (F) at day 0 and this area is filled by tissue engineered bioceramic block in other test groups that looked relatively denser than the adjacent bone. In group A, at day 14, there is no evidence of callus formation but there is increase in radiodensity of the construct with little periosteal reaction at the cut edges of the host bone. It is observed that there is mild resorption of the construct which started at day 30 and at the end of day 60 the resorption is moderate. There is close attachment of the autogenous construct with the adjacent ulna and bridging between construct and cut end of the host bone at day 60. Whereas, excellent new bone formation is noticed and bridging of construct with host bone and adjacent ulna is complete at day 90. In group B, bridging of the bone and construct at cut ends is seen from day 30 onwards. At day 60, there is complete bridging of composite allogenic construct with the host bone at all the surface of contact along with adjacent ulna which indicates the formation of new bone. At day 90, excellent new bone formation, which is evident radiographically by resorption of bioceramic block and bridging of construct completely with the host bone, and bioceramic block, have lost its shape completely.

The bridging between host bone and xenogenic construct in both proximal and distal interfaces is evident at day 30 in groups C and D. At day 60, there is close union between composite scaffold and the host bone as well as with adjacent ulna. A major portion of the construct underwent resorption and the boundary between construct and host bone disappear which indicates the construct have been mostly replaced by new bone at the end of 90 days. In groups C and D there is complete bridging on both proximal and distal ends with host bone of uniform radiodensity, which indicates deposition of new osteoid bone into the bioceramic construct by the osteoprogenitor cells. The bone formation is more or less similar in both C and D groups, however, bone formation in group D is better than group C. In animals of groups C and D there are no graft rejection process, which is radiographically evident by absence of inflammatory response in and around the critical size defect area. At this stage both the xenogenic transplanted groups (C and D), shows better bone formation compared to control (F) and ceramic group (E), but there are less bone formation in comparison to autologous (A) and allogenic group (B). In group E, there is gap between the bioscaffold and the cut ends of the host bone till day 90. There is no evidence of bridging at either proximal or distal end of the ceramic scaffold with the host bone. New bone formation is very negligible. In control group (F), at day 30, there is fibrous dense tissue appearance near the defect margins. The defect is clearly visible without any callus formation. At day 90, more fibrous tissue at the cut ends of radius, but there is no union of the ends by any of the structure. The total radiographic score for healing of critical size defect in different groups over a period of 90 days is significantly higher in group A followed by groups B, D, C, E and F.

The histopathological observations at different time intervals in various groups are presented in figure below. In group A, the periosteum formation is normal and completely surrounded the scaffold. Defect is completely filled with bony trabeculae which is mature at the junction and immature at the centre of the defect. Most of the constructs are substituted by newly formed bone with residual scaffold material that is visible at the defect site. The stained tissue section shows varying colors of pink with mineralized matrix seen as bright pink and the cell nuclei stained dark purple. At higher magnification, osteoblastic lineage is clearly observed around the new bone with many osteocytes. In group B, newly formed bone is observed in all the animals at defect sites and the bioceramic scaffold is completely surrounded by periosteum. The trabecular bone is observed throughout the tissue engineered construct, which is more mature at the junction of the defect compared to centre of the bone defect. Most of the scaffold is substituted by new bone with minimal residual ceramic material. At the junction of the defect site, there is higher amount of mature bone formation. The gap between the trabecular bones is minimal. Lamellar bone formation is seen into porous scaffold at higher magnification.

In groups C and D, the tissue engineered construct/bioscaffold is surrounded by the normal periosteum in all the animals. The new bony trabeculae are observed throughout the bioceramic construct. In higher magnification at the junction of the construct, the lamellar bone formation is seen along with proliferating osteoprogenitor cells. The residual material is minimal indicating resorption of ceramic material in these groups. The new bone stained light pink and ceramic material have a shadowy white appearance, loose connective tissue stained light pink and cells stained dark purple. There are no signs of inflammatory reaction around the construct in any of the animals of the groups C and D. In group E, the periosteal reaction surrounding the scaffold is minimal and formation of trabecular bone is very less. Most of the scaffold is filled with fibrous connective tissue which stained light pink colour throughout the scaffold. No phagocytes are observed in the evaluated sections indicates high biocompatibility of HA/TCP ceramic. In group F, there is no evidence of bone formation in any of the animals at the defect site. There is proliferation of dense fibrovascular connective tissue and inflammatory reaction at the junction of the defect. The entire gap is filled with fibrovascular tissue. The total histological scores of bone healing for different groups are presented in below Table. Group A have the highest scores, followed by groups B, D, C, E and F.

The scanning electron microscopic observations are presented in figure below. A close contact without intervening soft tissue is seen between the bones and construct in the group A. The osseous tissue is predominantly seen at the junction of construct and host bone. There is presence of bony tissues inside macropores of ceramic construct throughout surface of the block. The micron-sized micropores inside the block are infiltrated by bony tissue and the collagen fibrils demonstrated a direct bone-to-material contact. In group B, there is strong bridging of construct with the host bone by mature osseous bony structure. The bone formation is seen throughout the surface of ceramic-cell construct inside the micropores. Collagen fibrils are arranged in irregular manner and their density is more. In groups C and D, bioceramic construct is in close contact with the host bone by newly formed osseous tissue. The entire macropores of the bioceramic construct is filled with woven bone. The resorption of hydroxyapatite, growth of new bone from inside to outside in the macropores of ceramic construct along with formation of collagen fibers is also evident. In group E, bridging of bioceramic construct with host bone is not so strong unlike other test groups. The surface of the bioceramic block reveals more of the implant material as such, which is indicating less resorption of ceramic material. Inside the macropores, new bone formation is scanty. In group F, there is no bone formation at the entire length of the defect. The defect is filled with fibrous tissues.

Large bone defects resulting from extensive trauma, non-union, delayed union or tumour resections are common clinical problems. Despite the usefulness that minimally invasive surgery and osteosynthesis have brought fracture management and bone healing, there are still many occasions where bone healing remains challenging. In these cases uses of mesenchymal stem cells from different origin may prove helpful and could promote bone regeneration in a large bone defect as observed in this study uses of mesenchymal stem cells from different origin accelerated bone healing.

Conclusion

In vivo experiments in rabbit CSD model confirmed that autogenous, allogenous and xenogenous BMSC seeded on bioscaffold promoted faster healing of critical size defects. Hence, it is suggest that BMSCs are suitable for bone formation in fracture healing and non-union.

Table: Radiographic grading scale for degree of new bone formation and critical sized defect (CSD) healing

Description	Score
– No change from immediate post-operative appearance	0
– A slight increase in radio density distinguished from the implant	1
– Recognizable increase in radio density of the implant with bridging of the cortex	2
– Callus formation with complete defect bridging	3
– Bridging at least one cortex with material of non-union radio density, easily incorporation of the implant suggested by obscurity of implant borders	4
– Defect bridged on both proximal and distal sides with bone of uniform radio density, cut ends of the cortex still visible, implant and new bone not easy to differentiate	5
- Same as grade 3, with at least one of four cortices obscured by new bone	6
– Defect bridged by uniform new bone, cut ends of cortex no longer distinguishable, implant no longer visible	7
– Complete union, complete new bone formation, complete cortex covered the defect, marrow cavity formed	8

Table: Histological grading scale for degree of healing

Scale	Description	Score
Quality of Union	No sign of fibrous or other union	0
	Fibrous union	1
	Fibro cartilage or cartilage union	2
	Mineralized cartilage and immature bone union	3
Bone union		4

Table: Radiographic score between different groups in different time interval

Groups	Day 0	Day 14	Day 30	Day 60	Day 90	Total
Group A	0±0.00	1.2±0.09*	2.5±0.34	3.8±0.43*	4.5±0.45*	12.00±0.32**[a]
Group B	0±0.00	0.9±0.45	2.9±0.22*	3.4±0.23*	4.02±0.05*	11.22±0.23**[b]
Group C	0±0.00	0.6±0.78	2.7±0.67	2.9±0.57	3.98±0.78*	10.18±0.70**[c]
Group D	0±0.00	0.7±0.87	2.9±0.56*	3.5±0.23 *	4.1±0.77*	11.20±0.60**[c]
Group E	0±0.00	0.3±0.12	1.45±0.66	2.1±0.45	2.2±0.12	06.05±0.33*[de]
Group F	0±0.00	0±0.00	0.2±0.22	0.3±0.12	0.44±0.89	00.94±0.24[d]

Table: Histological score between different groups at 90th day

Implanted group	Histological Score Average ± Standard Deviation
Group A	3.0±0.23[a]
Group B	2.7±0.65[b]
Group C	1.9±0.63[c]
Group D	2.2±0.32[c]
Group E	0.75±0.23[d]
Group F	0.2±0.01[d]

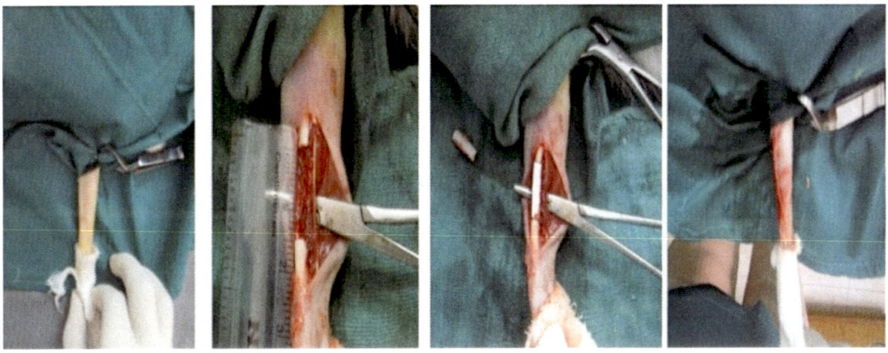

Fig. 1 : Surgical procedure

CSD = 15 mm in Radius of NZW Rabbit

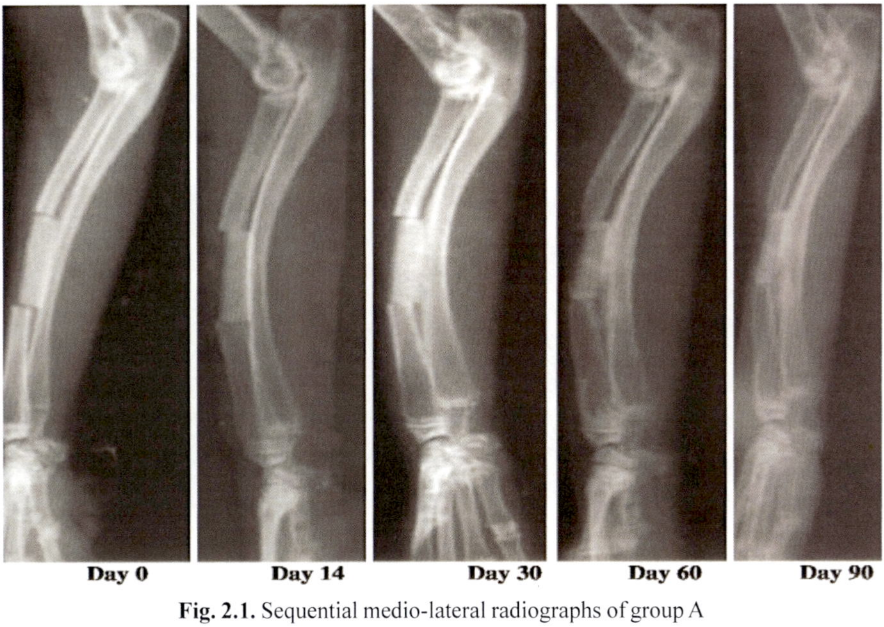

Fig. 2.1. Sequential medio-lateral radiographs of group A

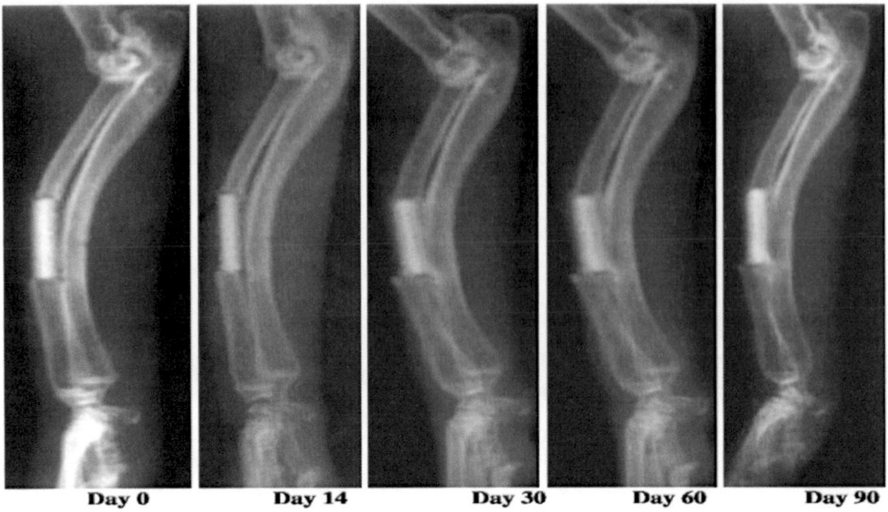

Fig. 2.2. Sequential medio-lateral radiographs of group B

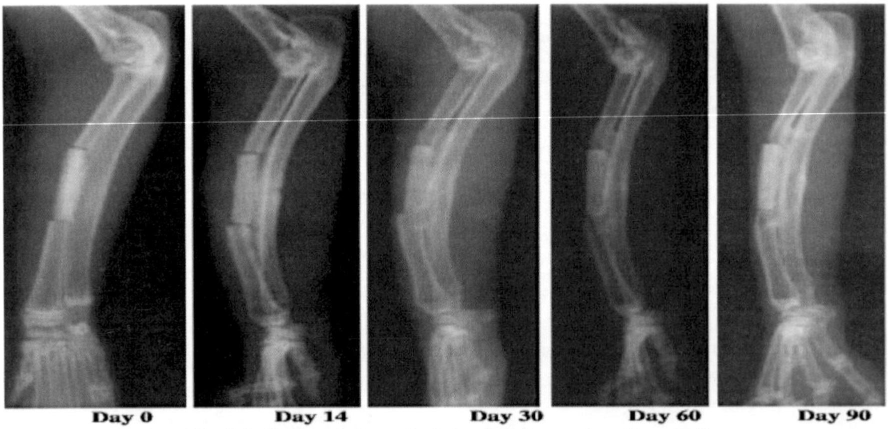

Fig. 2.3. Sequential medio-lateral radiographs of group C

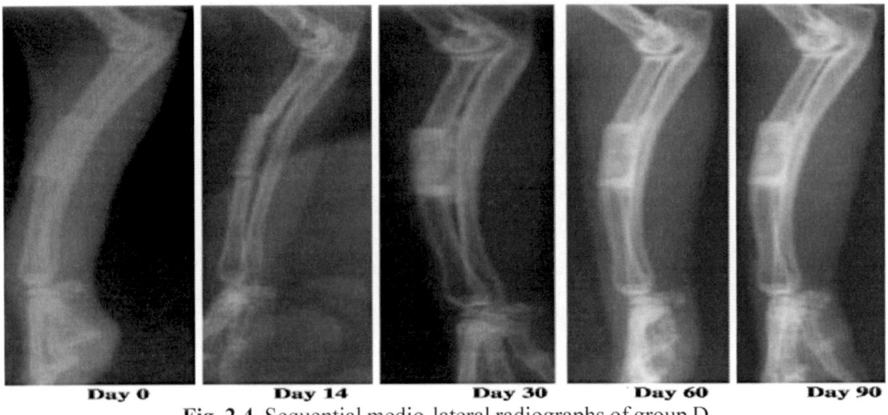

Fig. 2.4. Sequential medio-lateral radiographs of group D

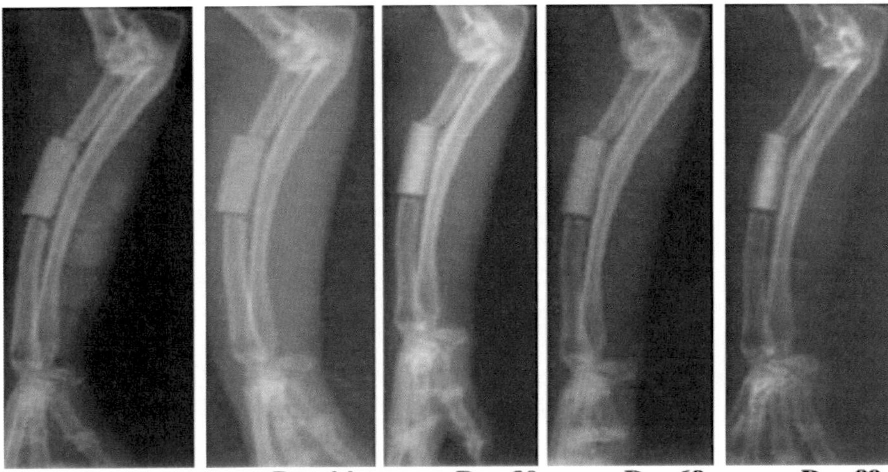

Day 0 Day 14 Day 30 Day 60 Day 90

Fig. 2.5. Sequential medio-lateral radiographs of group E

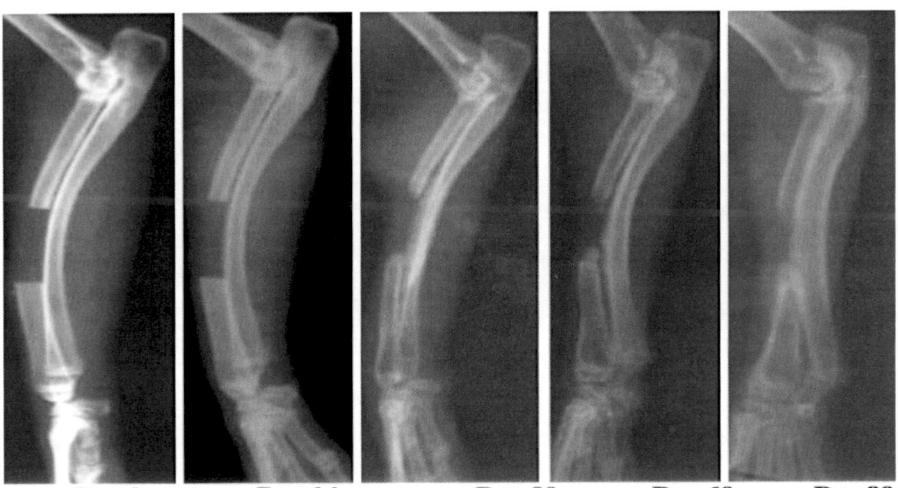

Day 0 Day 14 Day 30 Day 60 Day 90

Fig. 2.6. Sequential medio-lateral radiographs of group F

Fig. 2. Sequential radiographs of different groups at different time intervals

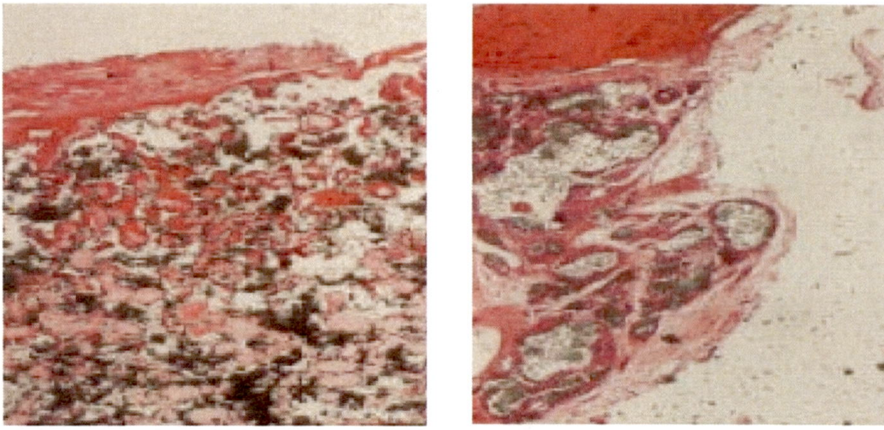

Group-A

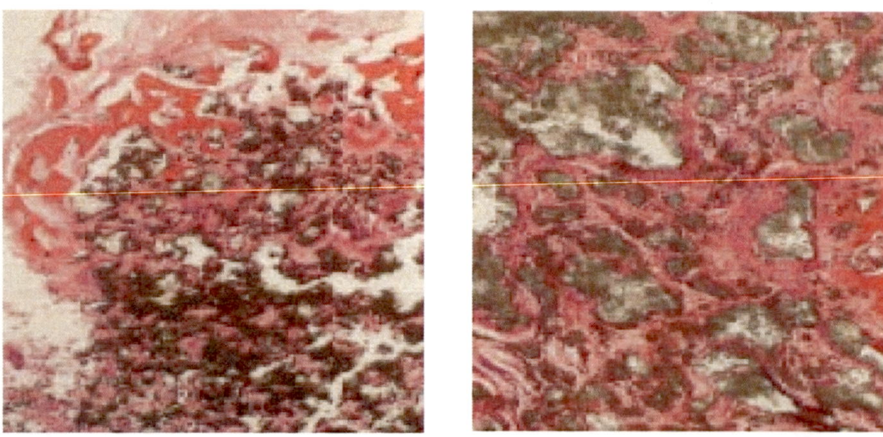

Group-B

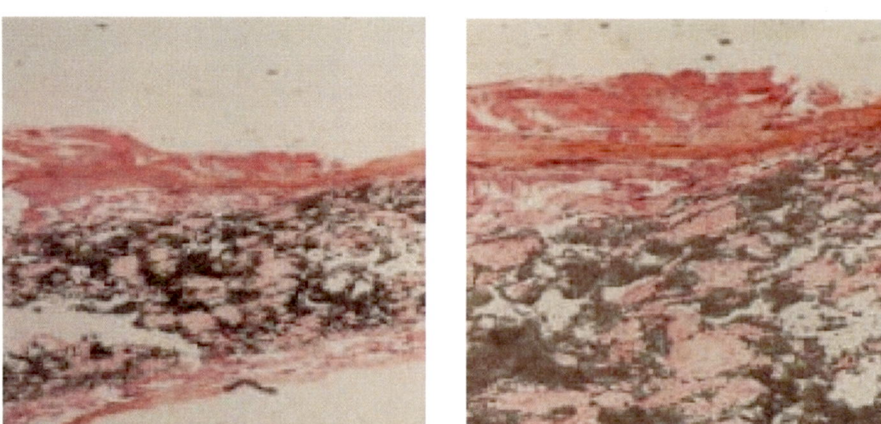

Group-C

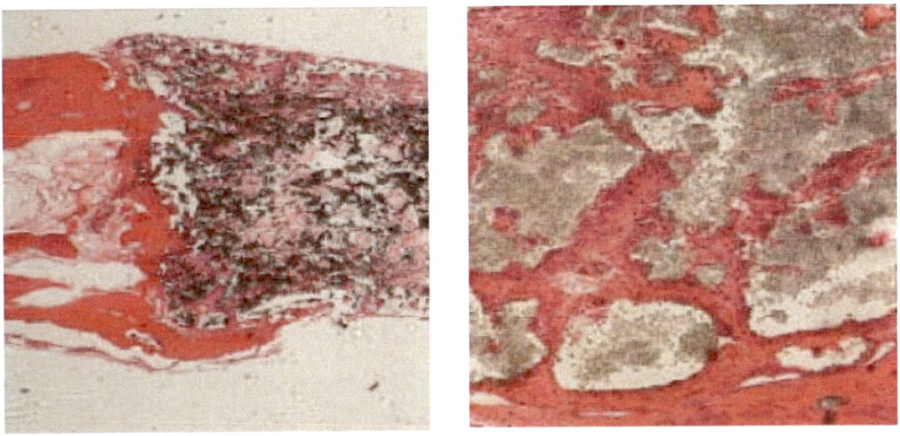

Group-D

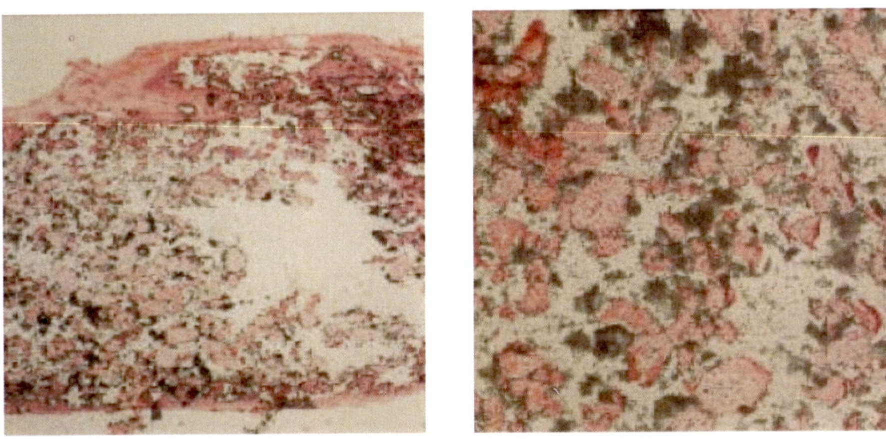

Group-E

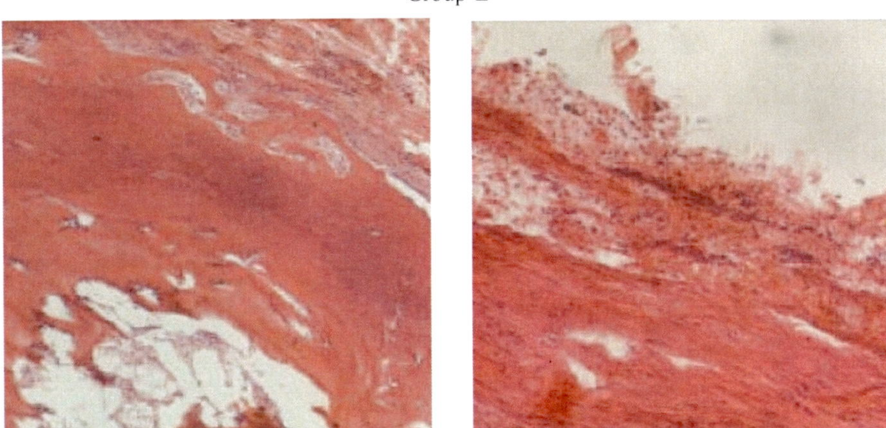

Group-F

Fig. 3. Photomicrograph of the healing CSD site of different groups

11

Mesenchymal Stem Cells Seeded Bioceramic Construct for Bone Regeneration in Large Critical-Size Bone Defect

Despite the benefits that minimally invasive surgery and osteosynthesis have brought to fracture management and bone healing, there are still many circumstances where bone healing remains challenging. Large bone defects are serious complications that are most commonly caused by extensive trauma, tumour, infection, or congenital musculoskeletal disorders. In nonunion cases repairing of bone defects with composite biomaterials as defect filler can promote bone regeneration. Currently, the gold standard for bone regeneration is use of autogenous bone graft. In order to avoid morbidity at the donor site or if large amount of autogenous bone is needed, bone substitution materials can be used. Bone substitution materials can be combined with cells such as mesenchymal stem cells (MSCs) to increase bone formation. Bone-marrow derived mesenchymal stem cells (BMSCs) represent an attractive cell population for tissue regeneration. Bone marrow stem cells are pluripotent cells that have been used to facilitate bone repair because of their capability of differentiating into osteoblasts. Several studies on the regeneration of bone have shown that cultured BMSCs, seeded on different bioabsorbable implants are able to induce bone formation *in vivo* and lead to improved healing of critical-size defects.

Calcium phosphate bone substitutes such as hydroxyapatite (HA) and tricalcium phosphate (TCP) are currently used for bone substitution in many different clinical applications such as repair of bone defects after trauma or tumour and bone augmentation in spinal arthrodesis. Although these bone substitutes are osteoconductive, they often lack the osteogenecity needed to support bone healing in large defects and are slowly degraded in the body. Studies have shown that coating of hydroxyapatite with a calcium silicate layer could encourage cell proliferation and osteogenic differentiation of human bone marrow-derived stromal cells. Silica-calcium phosphate composite in comparison to calcium phosphate-rich biomaterials has a faster resorption rate owing to greater dissolution of Si ions. The use of several growth factors have been studied in bone repair and these factors are known to play a role in differentiation of mesenchymal progenitor cells specific lineages like endothelial cells or osteoblast. Bone morphogenetic protein-2 (BMP-2) has been shown to accelerate bone healing in humans and animal models. BMP-2 acts by osteoinduction and is involved in the differentiation of mesenchymal progenitor cells into osteoblasts. The addition of recombinant human BMP (rhBMP-2) to a self-cross linkable cellulose hydrogels/biphasic calcium phosphate granules construct promotes bone regeneration in a critical-size segmental defect model of non-union in dogs. It is also reported that protein, insulin-like growth factor (IGF-1) stimulates direct migration of human mesenchymal progenitor cells (MPC) and contributes to the recruitment of MPC in bone formation and bone healing.

Transplantation of human mesenchymal stem cells in a non-autogenous setting for bone regeneration in a rabbit critical-size defect (15mm) has been reported in previous chapter in this book, but the regeneration potential of autogenous MSC with or without growth factors beyond standard critical-size defect is not reported. So, in this chapter, the potential of cultured autogenic rabbit bone marrow derived mesenchymal stem cells (rBMSC) seeded in a triphasic composite bio-scaffold (namely HASi) to generate new bone formation in an extensively large (30mm) segmental diaphyseal bone defect in rabbit model with or without growth factors rhBMP-2 and IGF-1 is discussed.

The study is conducted on 30 female adult New Zealand White rabbits (2.86± 0.38 kg, 6-9 months old) divided into five equal groups (each n=6). They are kept in individual cages, fed a standard diet and allow free mobilization during the study. The groups are compare as follows: The radial 30-mm segmental diaphyseal defect is filled with a HASi scaffold only (group B); HASi scaffold seeded with autogenous rBMSC (group C); HASi scaffold seeded with autogenous rBMSC along with rhBMP-2 (group D) and HASi scaffold seeded with autogenous rBMSC along with IGF-1 (group E). An empty defect serves as the control group (group A).

The same number of rBMSC (five millions cells) and concentration of growth factors rhBMP-2 (50μg) and IGF-1 (50μg) are again injected around the bone defect on day 15 postsurgery in their respective groups.

A triphasic composite bio-scaffold (calcium silicate, hydroxyapatite and tricalcium phosphate) namely HASi with elements in the following percentage: calcium-66.36%, Phosphorus- 25.35%, Silicon- 8.29% and porosity of 50 – 500 μm is used as ceramic block. Each HASi block is of 30 mm long and 5 mm in diameter are produced in an emulsion process.

Isolation and culture of rBMSC is performed as reported earlier (Maiti et al., (2012). For more detail: "Stem Cell Research in Lab Animals" (2020) by Swapan Kumar Maiti. The rBMSCs at passage 3 are used for this experiment.

The HASi bioscaffold are incubated at 4^0C in 50μM ml 1 fibronectin solution diluted in PBS for 24 h. Five million passage-3 rBMSC are resuspended in 4 ml of culture medium (DMEM) and transferred into a 5 ml tube. The HASi blocks are placed into the medium containing the cells. After 1.5 h of continuous spinning at 37^0C (35rpm), the ceramic blocks are placed into a 6-well plate. This MSC seeded ceramic scaffold is used in group C. Tissue engineered bone construct for group D consists of seeded HASi scaffold along with 50μg of rhBMP-2 suspended in 250 μl of DMEM, is injected at the implant site, whereas, in group E, seeded HASi scaffold along with 50μg of IGF-1 suspended in 250 μl of DMEM is injected at the implant site. In group B, only HASi block/scaffold is used at the implant site.

All the animals are acclimatized for two weeks to approaching and handling prior to surgery. Under general anaesthesia with xylazine hydrochloride (5mg/kg, IM) and ketamine hydrochloride (50mg/kg, IM) and strict surgical asepsis, a unilaterally critical-size bone defects are created in the radial diaphysis. A 4 cm super medial incision over the radius is given, soft tissue is dissected and the bone is exposed by gentle retraction of the muscles. A Hohmann retractor is placed between ulna and radius to protect the ulna. A 30 mm segmental osteo-periosteal diaphyseal defect is created with an oscillating bone saw that is continuously cooled by irrigation with 0.9% sterile cold saline solution. The periosteum is removed with the bone and 5mm of periosteum is stripped from each side of the remaining radius. The bone gap is irrigated with sterile physiological saline solution and ceramic block (or nothing) fitted into the gap as per treatment protocol for different groups. Muscles, fascia and skin are separately closed over the defect with 4-0 resorbable sutures. A wooden splint with bandage is applied on the test bone for one week.

Postoperatively, ceftriaxone sodium (22mg/kg, IM) is administered intramuscularly for five days. Analgesic meloxicam (0.2 mg/kg, IM) is given for the first 3 postoperative days. There are no differences in quantity of meloxicam administered between the groups. Water and food are supplied *ad libitum*. After 90 days of experiment, the animals are sacrificed.

Craniocaudal and mediolateral radiographic view of each ostectomy site is assessed immediately after surgery to monitor the position of the implants, and at 7, 15, 30, 45, 60 and 90 days postoperatively to assess bone formation and bone union with the host bone at both proximal and distal junctions. The radiographs are graded with slight modification of the scoring system reported by Yang *et al.* (1994). For each category mean scores are calculated. These mean scores of each time interval are added separately for obtaining the total radiographic scores. Overall radiographic scores for each group are calculated by adding mean scores for each group at different intervals. Resorption of the implant is assessed qualitatively.

Microcomputed tomography (μCT) technique is used for evaluation of three dimensional (3-D) trabecular new bone formation in different treatment groups. At day 90 after implantation, the constructs are explanted after euthanizing the animals, preserving both proximal and distal interfaces. Images are acquired under 70 keV voltage with a constant 114 μA current. Spatial resolution is limited to 100-200 μm. A high- resolution protocol (slice thickness 120 μm, feed 60 μm, and pixel size 60 μm) is applied. An X-ray tomographic microscope, Micro-CT 40 is used.

At day 90 post-implantation, the animals are sacrificed. The test bone sections throughout the length of the defect are collected from the site of surgical implant and washed thoroughly with sterile normal saline solution. Fixation is done in 10% formalin. The bone sections (longitudinal as well as transverse) are subjected to decalcification in Goodling and Stewart solution (15ml formic acid, 5ml formalin and 80ml distilled water). The decalcified bone sections are stained by H & E and Masson's trichrome stain. The sections are examined using a light microscope under different magnification. Bone healing is assessed in each group according to standard published paper / scoring system.

The food and water intake is returned normal on day 3 in all experimental groups (B, C, D, E) whereas, it is on 6-8 days in control group. Weight bearing on the operated limb is mild to moderate from day 3 after surgery in all groups except group A. Complete weight bearing is seen after removal of wooden splint in all animals of groups B, C, D and E on day 7 after surgery. In the control group the animal does not bear weight normally throughout the experiment period. In three animals of group A, (empty control), a fracture of the treated leg is observed, so these three animals are sacrificed prior to the study endpoint. No animal of this group A is survived at day 90. The implanted biomaterials do not cause any apparent signs of irritation or infection and cutaneous wound healing is uneventful.

The radiographic scoring is done based on mediolateral view as there is overlapping of the radius by ulna in craniocaudal view. Radiographic scores (mean ± SD) of construct appearance, osteotomy line, periosteal reaction, total and overall radiographic scores in all five groups at various time intervals are presented in below table. Mediolateral views of radiograph of animals of five groups are also presented in below figures.

In control group (A), at day 30, the osteotomy line at defect site become invisible marking the progress of bone repair near this cut end. At day 45, a slight increase in radio-opacity is seen near this defect, marking the beginning of callus formation.

On day 60, the area of radio-opacity is seen to be spread a little more and the defect remaining unbridged. No animal of control group is survived at day 90 as there is complete fracture of remaining ulna caused complete lameness and unable to stand or walk. In group B, mild resorption of the construct starts at day 45 and at day 60, the resorption is moderate. There is gap between the construct and the cut ends of host bone till day 60 but there is increased bone density at the junction of construct with the ulna at day 90. There is no evidence of bridging at either proximal or distal end of ceramic block with the host bone. Osteogenic process is very negligible. In group C, bridging of the bone and construct at proximal end is seen from day 30 onwards. At day 60, there is close attachment of the construct with the adjacent ulna and bridging between composite construct and cut end of the host bone. At day 90, good amount of new bone formation is noticed and the construct also underwent moderate resorption by this time. In group D, bridging between host bone and implant in both proximal and distal interfaces is evident at day 30. At day 45, there is close union between composite scaffold and the adjacent ulna. At day 90, a major portion of the construct underwent resorption and the boundary between implant and normal bone is disappeared, which indicates that the implant had been replaced by early new bone. In group E, there is bony union between scaffold and ulna on day 45, whereas, bridging of both the proximal and distal defect ends with the host bone is seen at day 60. Moderate resorption of construct and bridging with adjacent ulna is more evident at day 90. Statistical analysis of total radiographic score at day 15 shows that group D is significantly different from other groups. On day 60, there is significant difference between group C and group E. On day 90, all groups show significant variation from each other. Overall radiographic score show that group D have the highest score, followed by groups E, C, B and A.

Micro computed tomography assessment is able to distinguish between the three phases present inside the defects: newly formed bone, remaining triphasic bioceramic block and non-mineralized tissues. The μCT scanned slices of bone biopsy from different groups are presented in below figures. The transverse and longitudinal sections reveal the unbridged cut ends of the host radius and there is no osteogenic reaction in-between radius and adjacent ulna in any of the animal group A. In other groups, newly formed trabecular bone at the defect site in radius bone is seen. The HASi component in defect site is almost intact in group B, whereas it is in the process of resorption in groups C, D and E. In group C, signs of osteogenesis and process of remodeling are observed. In group D, transverse and longitudinal bone sections show the establishment of periosteal continuity in between the cut ends of old host bone (radius) and continuation of medullary cavity at the defect site. Newly formed trabecular bone is under remodeling process in groups D and E.

In control group (A) majority of the defect area is occupied by fibrous tissue. In group B, very little new woven bone formation, which is discontinuous in nature, is seen at the junction of scaffold and cut ends of the host bone. The scaffold is in the process of resorption and localized regions of new bone formation are seen as dispersed islands. In group C, new woven bone formation at the junction of scaffold and cut ends of host bone and osteochondral union between the ulna and scaffold is seen. There is invasion of periosteum from the adjacent ulna into the defect area. The scaffold is under active resorption and replaced by newly formed woven bone at some places. In group D, abundant new dense bone tissue formation and close bony union of scaffold with the host bone at both cut ends as well as with the adjacent ulna is seen. Scaffold is under active resorption and many places it is replaced by newly formed woven bone. Increased vascularisation near the junction between scaffold and the adjacent ulna is seen. Marrow formation is also evident in more than half of the implant area indicative of remodeling. Osteogenesis process is of intramembraneous type without any intermediate cartilage formation. Masson's trichrome staining reveal close association of newly formed bone in the defect area with the host bone represented by more areas of mineralization (blue staining) and new osteoid (deep red staining) tissue. In group E, osteochondral union is seen between the scaffold and cut ends of host bone as well as with the ulna. The scaffold has undergone cartilaginous transition and areas of new bone formation are evident. The total histological scores of bone healing for different groups are as group D have the highest score, followed by groups E, C, B and A.

Conclusion

The findings of this study shows that *in vitro* seeding of autogenous rBMSC on bioscaffold HASi could hasten the osteoconductive properties of scaffold and can be alternative to autogenous bone graft in the treatment of large bone defects and non-union. Addition of growth proteins particularly rhBMP-2 to autogenous MSC seeded HASi bioceramic construct could accelerate bone regeneration in a large critical-size-bone defect model controlled study.

Table: Radiographic grading scale for analyzing the bone healing

Grading category	Score
Periosteal reaction	
None	0
Minimal (localized to the gaps)	1
Medium (extends over the gaps or towards ulna)	2
Moderate (1/4 to 1/2 of the defect area)	3
Full (1/2 to full length of defect)	4
Osteotomy line	

Both the osteotomy lines completely radiolucent	0
One of the osteotomy lines partially radiolucent	1
Both the osteotomy lines partially radiolucent	2
One of the osteotomy lines invisible	3
Both the osteotomy lines invisible	4
Construct appearance	
Unchanged/intact	0
Mild resorption (one localized area)	1
Moderate resorption (more than one localized area)	2
Mostly replaced	3
Fully replaced	4

Table. Histopathological scoring system

S. No	Feature	Score
1.	**Osteogenesis**	
	No osteogenesis	0
	Weak osteogenesis	1
	Medium osteogenesis	2
	Good osteogenesis	3
	Perfect osteogenesis	4
2.	**Union**	
	No evidence of union	0
	Fibrous union	1
	Osteochondral union	2
	Bony union	3
	Complete organization of graft	4
3.	**Marrow**	
	None in the resected area	0
	Beginning to appear	1
	Present in greater than 1/2 defect	2
	Complete colonization by red marrow	3
	Mature fatty marrow	4
4.	**Cancellous bone/medullary bone**	
	No osseous cellular activity	0
	Early apposition of new bone	1
	Active apposition of new bone	2
	Reorganizing cancellous bone	3
	Complete re-organization of cancellous bone	4
5.	**Cortical/compact bone**	
	None	0
	Early appearance	1
	Formation underway	2
	Mostly re-organized	3
	Completely formed	4

Table: Radiographic total scores of various parameters in five groups (A-E); Time intervals (7, 15, 30, 60 and 90 days) and their interactions.

Group and day	Construct appearance	Osteotomy line	Periosteal reaction	Total score
A×7	No construct	0.00±0.00	0.00±0.00	0.00±0.00[a]
B×7	0.00±0.00	0.00±0.00	0.00±0.00	0.00±0.00[a]
C×7	0.00±0.00	0.00±0.00	0.00±0.00	0.00±0.00[a]
D×7	0.00±0.00	0.00±0.00	0.00±0.00	0.00±0.00[a]
E×7	0.00±0.00	0.00±0.00	0.00±0.00	0.00±0.00[a]
A×15	No construct	0.00±0.00	0.00±0.00	0.00±0.00[a]
B×15	0.00±0.00	0.00±0.00	0.00±0.00	0.00±0.00[a]
C×15	0.00±0.00	0.50±0.55	0.33±0.52	0.83±0.02[c]
D×15	1.67±0.52	1.67±0.52	0.50±0.55	3.84±0.02[d]
E×15	0.00±0.00	0.67±0.52	0.17±0.41	0.84±0.02[c]
A×30	No construct	1.17±0.41	0.00±0.00	1.17±0.41[a]
B×30	0.65±0.55	0.85±0.52	0.00±0.00	1.50±0.02[a]
C×30	1.00±0.63	1.50±0.55	0.67±1.03	3.17±0.26[c]
D×30	2.50±0.55	3.17±1.33	1.50±0.55	7.17±0.45[d]
E×30	1.50±0.55	1.50±0.55	1.67±1.37	4.67±0.47[c]
A×60	No construct	1.67±0.52	0.00±0.00	1.67±0.52[a]
B×60	1.50±0.55	1.33±0.52	0.55±0.55	3.38±0.02[a]
C×60	1.67±0.52	2.67±0.52	2.33±0.82	6.67±0.17[c]
D×60	2.67±0.52	3.50±0.84	3.33±1.03	9.50±0.26[d]
E×60	2.67±0.52	3.00±0.89	2.00±1.26	7.67±0.17[e]
A×90	No construct	2.00±0.00	0.67±0.82	2.67±0.82[a]
B×90	1.83±0.41	2.33±0.52	1.67±0.52	5.83±0.06[b]
C×90	2.00±0.63	3.50±0.55	3.33±0.82	8.83±0.14[c]
D×90	3.50±0.84	3.83±0.41	3.33±0.03	10.66±0.23[d]
E×90	2.83±0.00	3.17±0.75	3.83±0.41	9.83±0.03[e]

Table. The total histological scores of bone healing in different groups

Group	Osteogenesis	Union	Marrow	Cancellous/ medullary bone	Cortical/ compact bone	Total score
A	1.00±0.00	1.33±0.00	0.00±0.00	0.00±0.00	0.00±0.00	2.33±0[a]
B	2.00±0.00	1.33±1.15	0.67±0.58	0.67±0.58	0.00±0.00	4.67±0.33[b]
C	1.67±0.58	2.66±0.58	0.33±0.58	1.33±0.58	0.00±0.00	5.99±0.21[c]
D	3.00±0.00	3.00±0.00	2.00±0.00	3.00±0.00	1.66±0.58	12.66±0.58[d]
E	2.33±0.58	2.33±0.58	1.00±0.00	2.00±0.00	0.00±0.00	7.66±0.33[e]

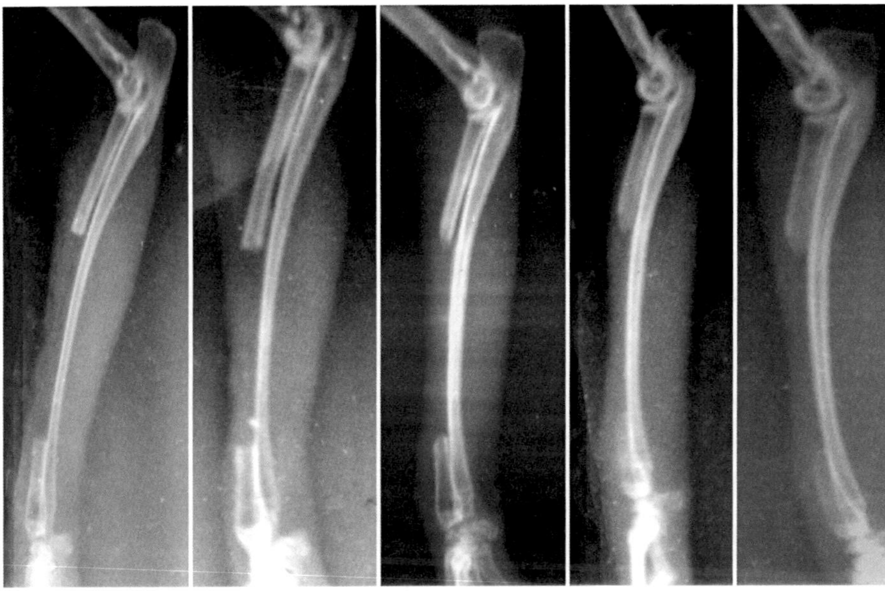

Day 7 15 30 45 60

Figure: Mediolateral radiographs of group A at different day intervals

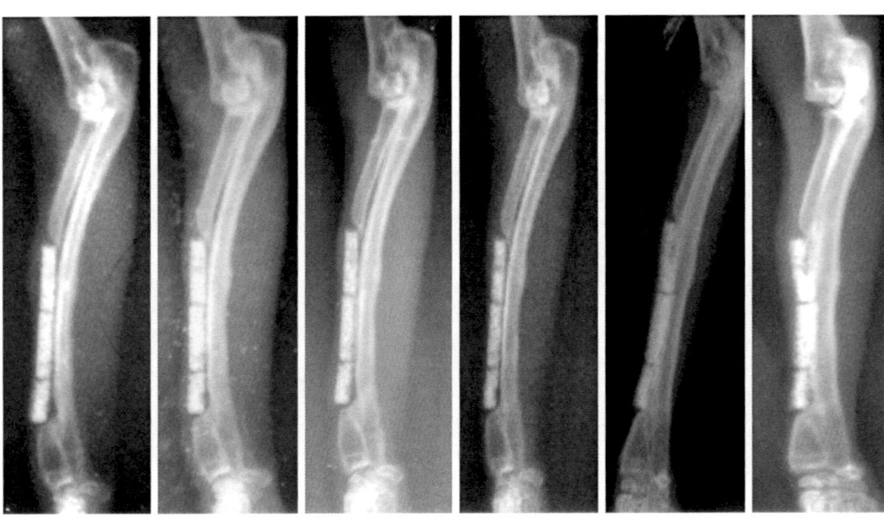

Day 7 15 30 45 60 90

Figure: Mediolateral radiographs of group B at different day intervals

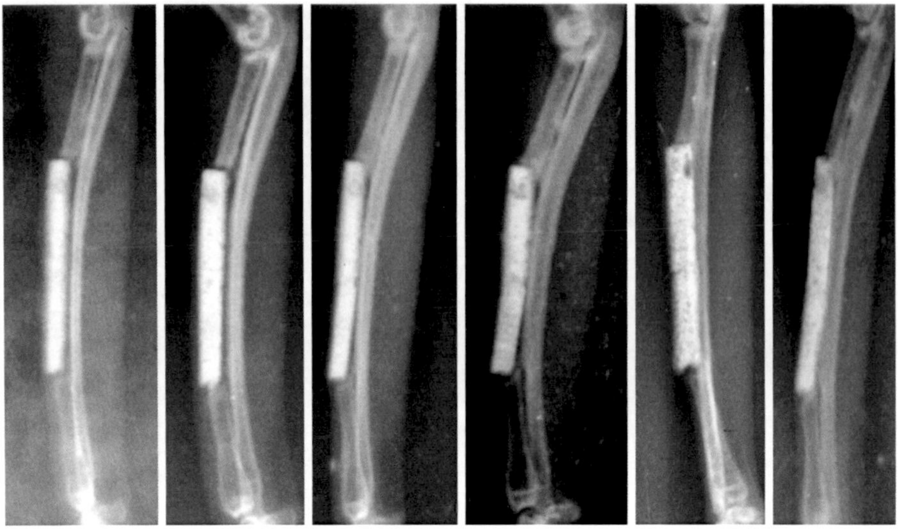

Day 7 15 30 45 60 90

Figure: Mediolateral radiographs of group C at different day intervals

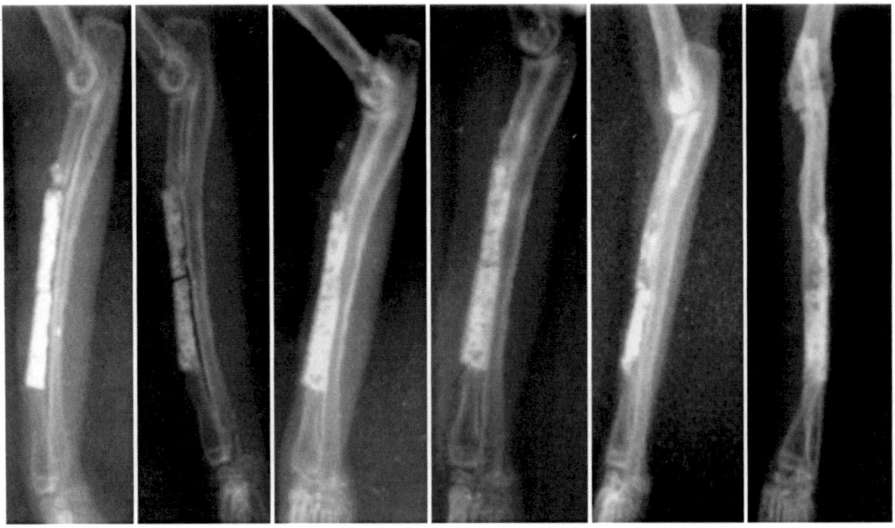

Day 7 15 30 45 60 90

Figure: Mediolateral radiographs of group D at different day intervals

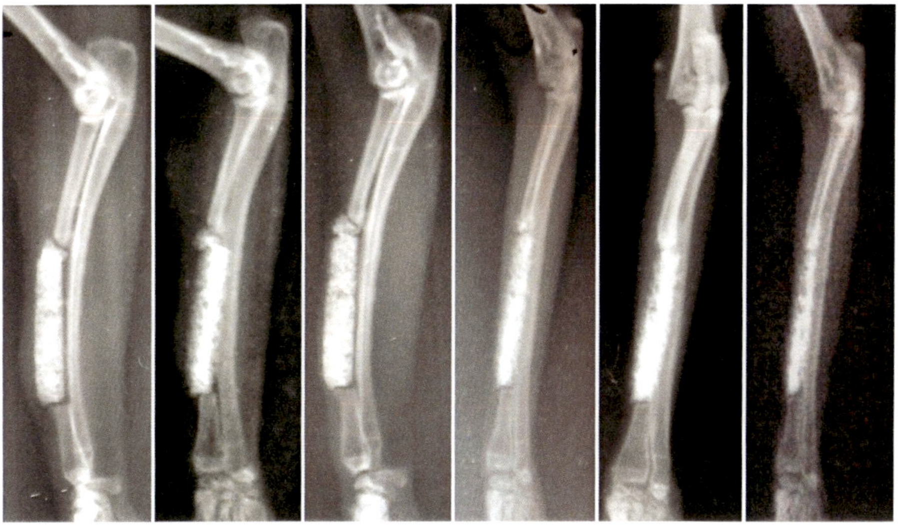

Day 7 15 30 45 60 90

Figure: Mediolateral radiographs of group E at different day intervals

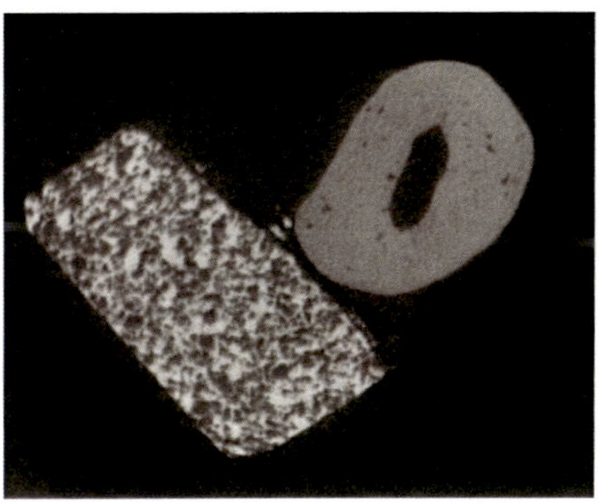

Figure: 3D Transverse bone biopsy slice from group B at day 90

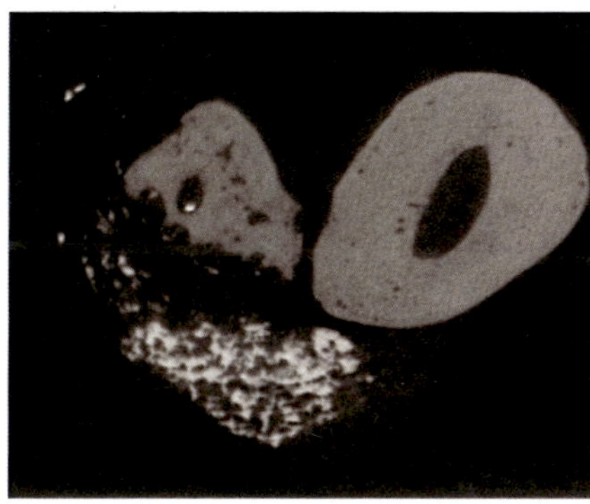

Figure: 3D Transverse bone biopsy slice from group C at day 90

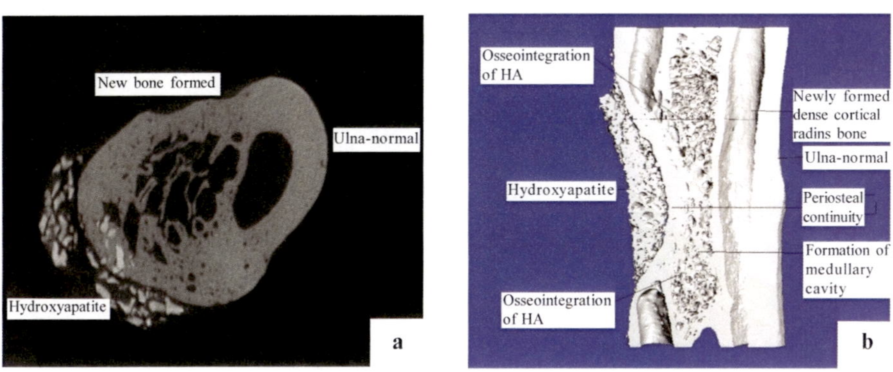

Figure: 3D (a&b)Transverse and longitudinal bone biopsy slice from group D at day 90

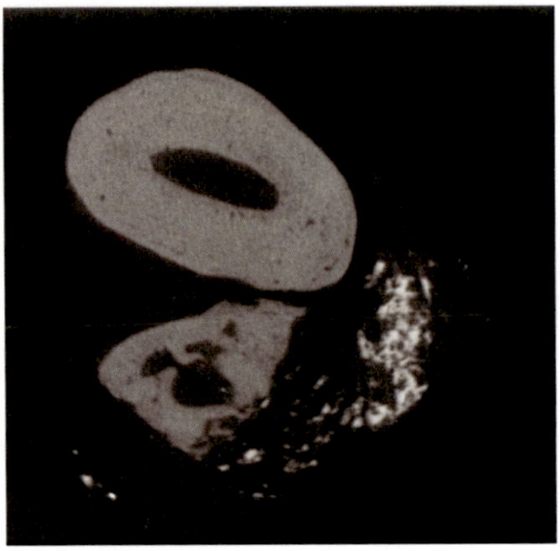

Figure: Transverse bone biopsy slice from group E at day 90

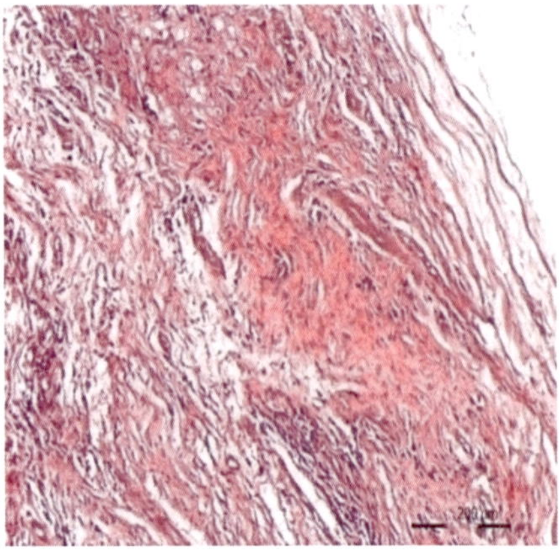

Figure: Longitudinal section of bone at defect site of group A shows fibrous tissue profilearation at the cut ends, H&E

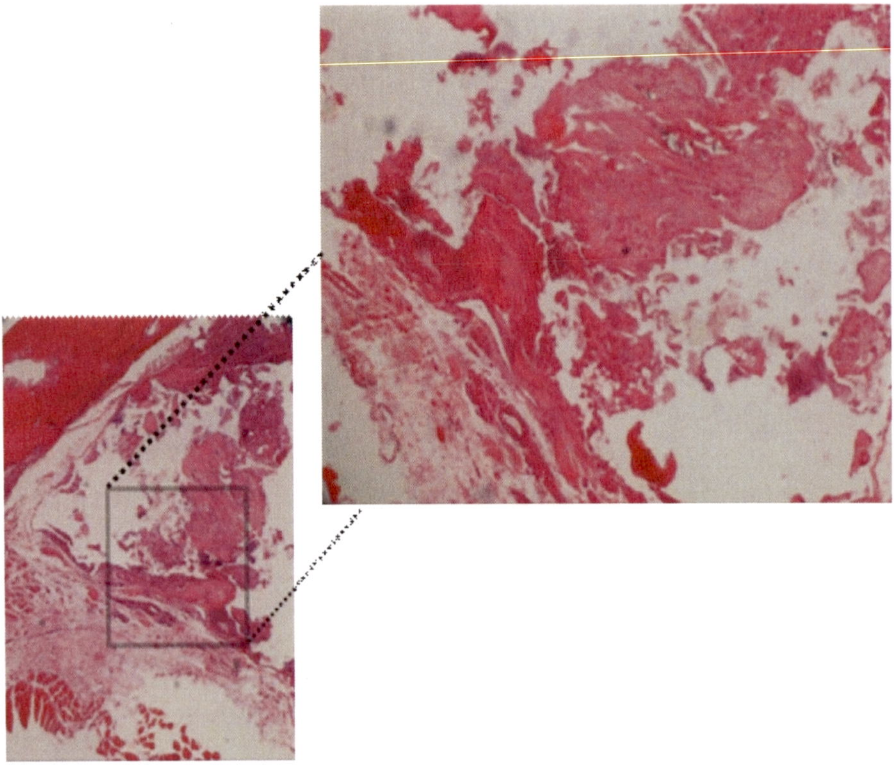

Figure: Transverse section of bone sample of group B shows few areas of new bone formation at the junction of scaffold and host bone, H&E

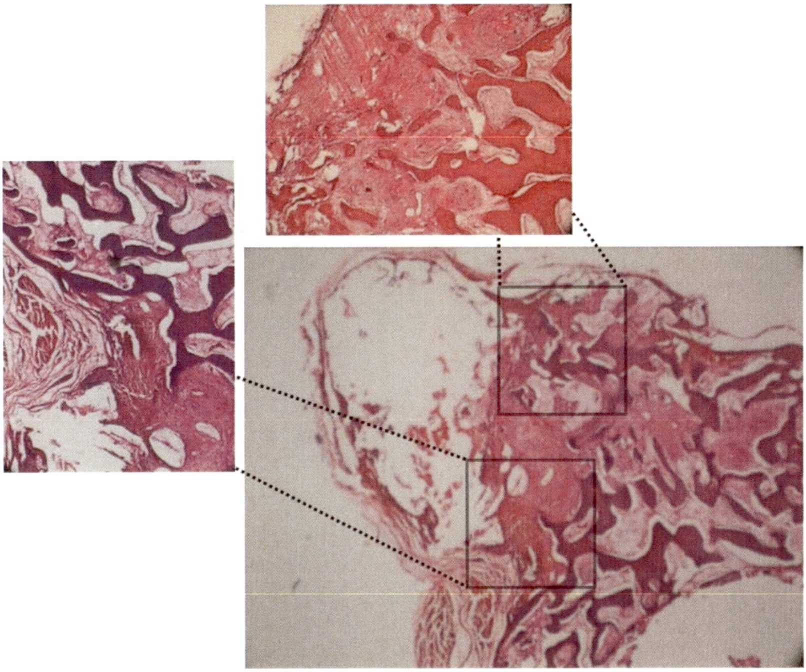

Figure: Transverse section of bone sample of group C shows osteochondral union between the ulna and scaffold, H&E

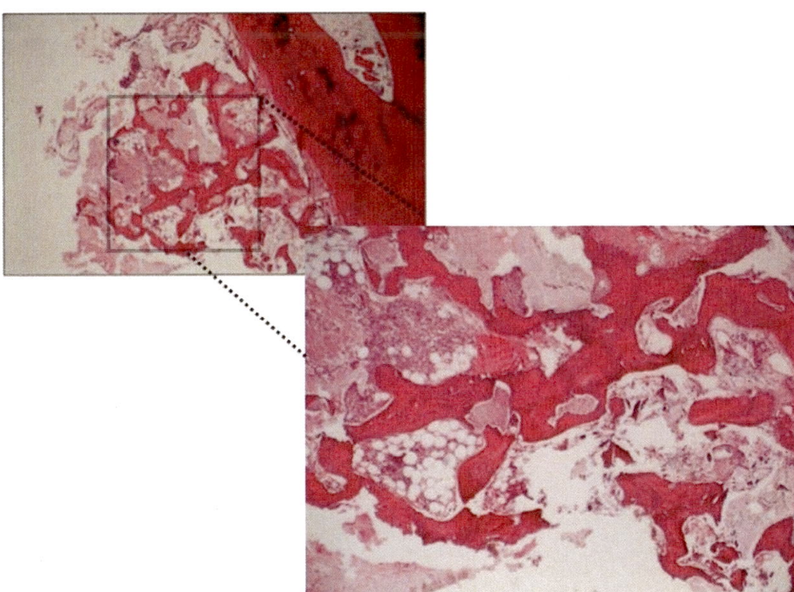

Figure: Transverse section of bone sample of group D shows extensive bony union of scaffold with the ulna, H&E

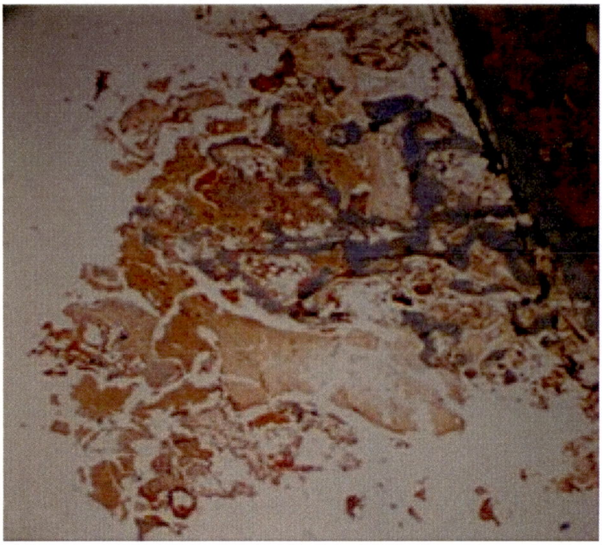

Figure: Transverse section of bone sample of group D shows newly formed bone in the defect area repreasented by areas of mineralization (blue staining) and new osteoid (deep red staining) Masson's Trichome

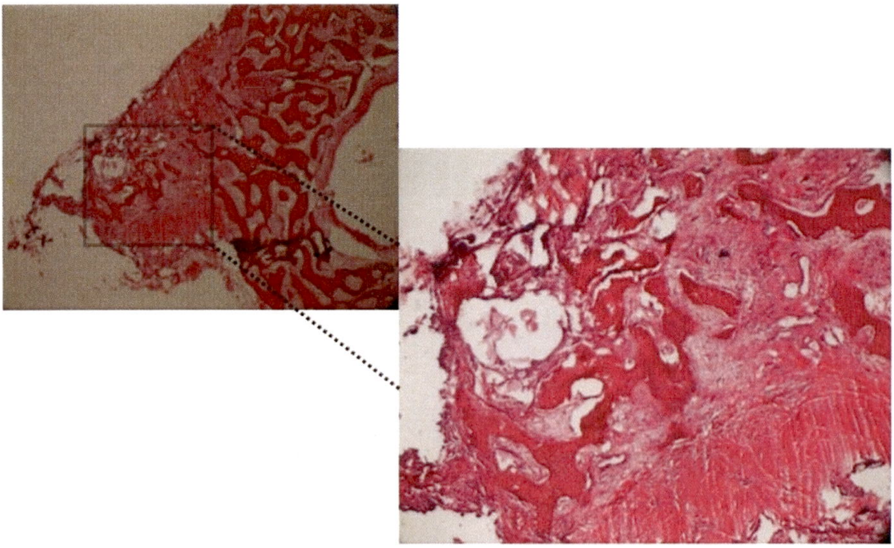

Figure: Transverse section of bone samples of group E shows the osteochondral union between the ulna and scaffold

12

Embryonic Stem Cell (ESC) Hydroxyapatite Construct with Growth Proteins for the Repair of Bone Defects

Bone is the only tissue in the body able to heal without microscopical scarring. On the other hand fracture healing is a slow process causing long immobilization periods with consequently high costs for fracture healing giving rise to delayed healing, fibrous healing or non-healing with subsequent problems for both, the patient and the orthopedic surgeon. Fracture healing is complex phenomenon which involves numerous cells, regulators of cell function and biochemical interactions in the repair process. The healing potential of bone, in a fracture model is influenced by a variety of biochemical, biomechanical, cellular, hormonal and pathological mechanisms. A continuous occurring state of bone deposition, resorption and remodeling facilitates the healing process.

Biodegradable and bioinert ceramic materials such as hydroxyapatite (HA), tri-calcium phosphate (TCP), aluminum-calcium-phosphate (ALCAP) ceramics provides scaffold to support the attachment and migration of newly formed bone cells into the osseous defect and also help in formation of a vascular network through the newly formed bone. Bioceramic also act as a carrier to deliver stem cells for osteogenesis.

Specific bone-inductive proteins/growth factors can induce bone formation and healing *in vivo*. These morphogens are therefore, to be an ideal alternative to autogenous bone grafts. These findings initiating intensive research into bone regeneration orchestrated by putative soluble signals and lead to the discovery and identification of an entirely novel family *of* protein initiators collectively called the bone morphogenic protein (BMPs), which belong to transforming growth factors-ß (TGF- ß) super family. These growth factors might possess a biological activity that is site and tissue specific in the induction of bone formation. Different growth factors have been demonstrated to possess osteoinductive and fracture healing properties.

Stem cells have therapeutic potential in the realm of orthopaedic surgery because of their capacity to self-renew and differentiate into various types of mature cells and tissues, including bone. Embryonic stem cells (ESCs) are pluripotent cells derived from the inner cell mass of pre-implantation embryos and represent embryonic precursor cells that give rise to any cell type in the embryo. Specifically, it has been shown that ES cells can differentiate into osteogenic cells under selective culture conditions. These osteogenic cells are capable of *in vitro* producing bone nodules in three dimensional scaffolds.

The potentials for using osteogenic stem cells in orthopaedic for bone and cartilage defect healing is immense, and is likely to further expand significantly in future. These approaches have included the implantation of osteoconductive extracellular matrix scaffold and the implantation of bone morphogenic proteins in various matrices. The most novel approach for this purpose is to combine a scaffold-matrix, living cells and biologically active molecules or factors together to form a tissue engineered construct (TEC) to promote the repair and regeneration of bony tissue.

In this chapter, the potential of cultured allogenic embryonic stem cells seeded in a calcium hydroxyapatite (HA/TCP) bioscaffold to generate new bone formation in a segmental bone defect with or without growth factors in rabbit model is described.

The study is conducted in 36 male adult New Zealand White rabbits divided into six equal groups. The groups and treatment on each group are as follows:

Group	Treatment
A	Empty defect (Control)
B	Defect filled with HA-TCP bioceramic
C	rESC with HA-TCP bioceramic
D	rESC with HA-TCP bioceramic + BMP-2
E	rESC with HA-TCP bioceramic + TGF-ß1
F	rESC with HA-TCP bioceramic + IGF-1

A biphasic composite bioceramic (calcium hydroxyapatite and tricalcium phosphate) namely HA-TCP with elements in the following percentage: calcium hydroxyapatite-65% and tricalcium phosphate 35%, and porosity of 200–450 μm is used as sterile ceramic granules.

Isolation of rabbit embryonic stem cells

Embryo recovery and collection: Each female rabbit is injected with PMSG intramuscularly in normal saline. After 72 hours of first injection, female rabbits are injected with HCG intravenously. Female rabbits are mated naturally. Collection of embryos is done by flushing the uterine horns and oviducts of rabbits 84-96 hours after mating. Embryos are collected at blastocyst stage by flushing the uterine horn with modified phosphate buffered saline.

Derivation of ES Cells: The embryos collected are seeded on buffalo fetal fibroblast feeder layers. In this technique whole embryo, with its zona pellucida are plated on a mitomycin-C treated feeder layer. Two days after plating, the attachment of the whole embryo to the feeder layer is observed and the medium replaced with fresh medium on a daily basis. The trophoectoderm layer began to expand on the 2-3 day of culturing. The inner cell mass (ICM) like cells is formed dome structure. The cells are then incubated at 37°C in a humified 5% carbon dioxide environment in a CO_2 incubator. Once confluence, cells are passaged 2-3 times to increase the cell population. After ES cells population is established, it is then confirmed by characterization with different markers by alkaline phosphatase staining and immuno cytochemistry.

Preparation of fibroblast monolayer, collection of fetus, isolation of fetal skin fibroblast cells, setting of monolayer of fibroblast cells of fetal skin and sub culture is described in detail in published paper: Bhadane et al (2018).

All the animals are operated under general anaesthesia with xylazine hydrochloride and ketamine hydrochloride and strict surgical asepsis, a unilaterally 5mm bone defects are created in the radial diaphysis. A 2 cm medial incision over the radius is given, soft tissue is dissected and the bone is exposed by gentle retraction of the muscles. A Hohmann retractor is placed between ulna and radius to protect the ulna. A 5mm segmental osteo-periosteal diaphyseal defect is created with an oscillating bone saw that is continuously cooled by irrigation with 0.9% sterile cold saline solution. The periosteum is removed with the bone and 5mm of periosteum is stripped from each side of the remaining radius. The bone gap is irrigated with sterile physiological saline solution and ceramic granules (or nothing) fitted into the gap as per treatment protocol for different groups. Muscles, fascia and skin are separately closed over the defect with resorbable sutures. A wooden splint with bandage is applied on the test bone for one week.

All the animals are kept in isolated cages and administered with broad spectrum antibiotic for 5 consecutive days. The analgesic is administered for 3 days. The wound is dressed until the complete healing of the incision is seen and sutures are removed. The animals are observed for the evidence of wound healing, radiographic bone healing postoperatively and they are euthanized for the study of gross morphology, histology, and scanning electron microscopy at 40th postoperative day. Medio-lateral and anterio-posterior radiographs are taken immediately after implantation and subsequently on days 7, 15, 30, and 40th postoperative day

Superovulation is successfully achieved in female New Zealand rabbits following injection of PMSG-HCG protocol followed by natural mating. A very good quality and good numbers of embryos are collected by flushing the uterine horns and oviducts of rabbits after 84-96 hours of mating.

The medium is used for culture and derivation of the stem cells consisted of Knockout Dulbecco's modified Eagle's medium with 5% FBS and recombinant murine Leukemia Inhibitory Factor (LIF) in 5% CO_2 and 37°C.This medium is found to be very useful for culture and derivation of rESC. In this study the whole blastocyst are seeded on the inactivated feeder rather than mechanical isolation of ICM by immunosurgery. The blastocyst stage of embryos those are ready to hatch at the time of collection with bulging of zona pellucida at one point is hatched within 4-6 hours after culture. Embryos those do not show attachment even after of 2-3 days of hatching are discarded and not used for further study. The ICM is allowed to grow on feeder cells for 12-15 days before the first transfer of the growing cells to a new feeder plate. ICM derived outgrowths that are mechanically removed from feeder cells after the perimeter of the outgrowths folded back. The ESC like colonies are mechanically separated from trophoectoderm cells and divided into small clumps by needle, then treated in a drop exposure to 0.05% Trypsin -Versine solution for 1 minute at room temperature. To terminate digestion, the culture is rinsed once in PBS, resuspended in fresh culture medium, and mechanically disaggregated into small clumps. The small ICM cell clumps are seeded on fresh BFF and the ESC- like colonies are seen several days later. The initial outgrowth of the ICM is often contained heterogeneous cell types. After growing approximately for 6-8 days the colonies are again lifted and passaged further for more separation of cells followed by treatment with 0.05% trypsin-versine solution for few minutes and then propagated in clumps of approximately 500 or more cells on fresh BFF. Embryonic stem cells are expressed high levels of alkaline phosphatase (AP) activity as evidenced by its characteristic pink coloration after staining. Expression of pluripotent markers in cultured embryonic cells is observed by alkaline phosphatase is further confirmed by Oct-4 in few cases by immuno-cytochemistry as evidenced by its green fluorescence. Colonies those shows positive expression are only used for implantation study in segmental bone defect in rabbit model.

The combination of xylazine and ketamine anaesthesia is provided safe and satisfactory anaesthesia for collection of embryos and during the implantation in rabbits and in none of the animals complications due to anaesthesia are observed. The hydroxyapatite -tri calcium phosphate granules are used in the form of granules and are secured at the defect site by suturing surrounding soft tissues. The process of transplantation of embryonic stem cells and growth factors at the site of osteotomy is carried out easily without any complication. Suturing of muscles, soft tissues and skin separately is helped to secure the implants in position and prevented their migration in all animals. External splint application to the respective forelimb immediately after implant application is found very useful to prevent the chances of complete fracture occurrence at the defect site.

Clinically, all the animals of different groups behaved identically, surgical wounds are healed by first intention. All the animals of six groups show signs of lameness in the immediate postoperative period. However, there is no visible difference in lameness between different groups. Weight bearing is gradually improved as signs of inflammation subsided within 8-10 days.

The radiographic evaluation is done based on mediolateral view as there is overlapping of the radius by ulna in craniocaudal view. In control group (A), at day 30, a fibrous density appearance is seen near the defect margins and there is no indication of the filling of the defect with newly formed tissue. At day 40, radiological observations are more or less the same except that the newly formed osseous tissue at cut ends is more organized and the fractured ends became more smooth and round. In group B, at day 20, HA-TCP bioceramic granules is appeared relatively more compact with signs of osteogenesis from the periphery. The periosteal reaction is seen adjacent to the cut ends of radius. On day 40, the bone defect is partially filled with newly forming bone; however, the implant is still present. There are indications of slowing down of healing process as the edges appeared relatively smooth and newly formed bone is organized. In group C, at day 10 and 20, the defect site shows increased radiographic density in all animals of this group. The cut ends of the bone are looked hazy and irregular with signs of new bone formation on either ends. On day 30, a radiolucent line is present in the middle of the defect showing incomplete bridging of the gap. On day 40, radiographs show that there are further consolidation of new bone formation and signs of union between the implant and host bone is visible at the proximal end of segmental defect. The implant is looking more compact at this stage indicative of progression of good healing process. In group D, at day 30, there is further increase in new bone formation at the defect site. At day 40, all the animals show signs of increased new bone formation and periosteal reaction at defect site as compared to day 30; however, still there are radiolucent lines visible in between the implant. In group E, at day 20, the signs of new bone formation from both cut ends of radius and radiodensity are increased at the defect site reducing the gap. At day 30, the defect site is completely healed and fused with the adjacent host bone in 4 animals. In remaining animals, new bone growth is seen from bone ends and only a small radiolucent line is visible between the implant and cut end of bone at distal fragment. On day 40, the defect site is appeared more radiodense with complete union between the implant and bone at proximal end of defect. In group F, at day 10, periosteal reaction is started. At day 20, there is increased radiodensity at the defect site, which shows signs of new bone formation. Cartilaginous density is observed at the defect margins on day 30. On day 40 the whole defect shows cartilaginous density whereas the defect margins show bony density and bridging the defect ends.

Histological sections at bone defect site of different groups at day 40 are recorded. In control animals (A), majority of the defect area is occupied by fibrous tissue. However, there is no continuity between fibrous tissue from the proximal and distal fragments. In group B, the defect site shows merging of osteoid bone ingrowths in area adjacent to the fibrous connective tissue with tendency to form Haversian systems and also cavitations lined by many osteoblasts. There are signs of resorption of implanted material. In group C, the histological picture at the defect site shows only partial degradation and absorption of HA-TCP particles. There is no infiltration of any inflammatory cells. However, the connective tissue below the periosteum and underneath the implanted construct mass is moderately thick, having many fibroblasts and osteoblasts with tendency to form Haversian systems and also remodeling with cavitations which are generally lined with osteoblast and also appearance of marrow elements in some of the cavities. The bone defect is almost repaired with newly formed bone. In group D, thick periosteal connective tissue covering over the defect site which is seen invading in and around the implanted bioceramic, which is entrapped within the interlacing osteoid trabeculae. There is evidence of hypertrophied chondrocytes indicating the bone formation through endochondral ossification. The trabecular structures of the bone are lined by many osteoblast cells. In group E, thick band of fibrous connective tissue is seen invading the implanted material from all sides and the interface of the implant-host bone shows intense osteoblastic activity forming woven bony structures having many osteoblasts and few osteoclasts. The new bone formation is found to be comparatively better than groups B and C. In group F, mild to moderate amount of implanted material is getting embedded into mature bony tissue having developed Haversian system. The osteoid tissue is more prominent forming woven bone pattern and the stroma consisted of large number of osteoblasts and osteocytes. There is complete filling of the defect area by new bone growth. The interlacing bony trabeculae are seen covering grafted material at the defect site. The margins of the trabecular bone are lined by osteoblasts and osteoclasts cells.

SEM analyses at bone defect site of different groups at day 40 are recorded. In group A, the gap is filled by newly proliferated tissue which is a mixture of cartilaginous and fibrous tissues. The fibro cartilage is present at the centre of the defect. Adjacent periosteal surface of the ulna is also attached with healing site through newly proliferated tissues. However, it is only fibroblast that invaded the defect. In group B, HA-TCP particles are seen projecting over the surface.

These particles are seen surrounded by newly formed bone matrix suggesting that bioceramic provided base for new bone formation, however, it is failed to incorporate with newly formed bones. In group C, the healing process of new bone formation is relatively faster as compared to group B. The ceramic particles are seen fused with newly formed bone matrix suggesting their incorporation with newly formed osseous tissue. Unorganized osseous tissue is advanced from the cut edges into the implant material. The area of the implanted site is almost completely occupied by the bone matrix. The extracellular matrix is well organized at many places indicating maturity of the osseous tissue. In group D, the defect site reveals some irregular area in the centre and at the host implant junction there is presence of compact bone suggesting advanced stage of healing. In group E, the SEM picture depicts compact bone formation at the host and implant junction towards the periosteal surface of the bones. Further magnification of the implanted site shows well organized trabecular bone embedded in the implanted material which is surrounded by compact bone at places. In group F, scanning micrograph of healing site shows some irregular areas in the centre of the defect and at the junction of the bioceramic construct and the cut edges, there is presence of compact bone suggesting advance stage of healing.

Conclusion

Adding of allogenous rESC on bioscaffold HA-TCP can hasten the osteoconductive properties of bioceramic and can be alternative to autogenous bone graft in the treatment of delayed and non-union. Addition of growth proteins includes BMP-2, TGF-ß and IGF-1 to ESC seeded HA-TCP bioceramic construct could accelerate bone regeneration in a segmental bone defect model controlled study.

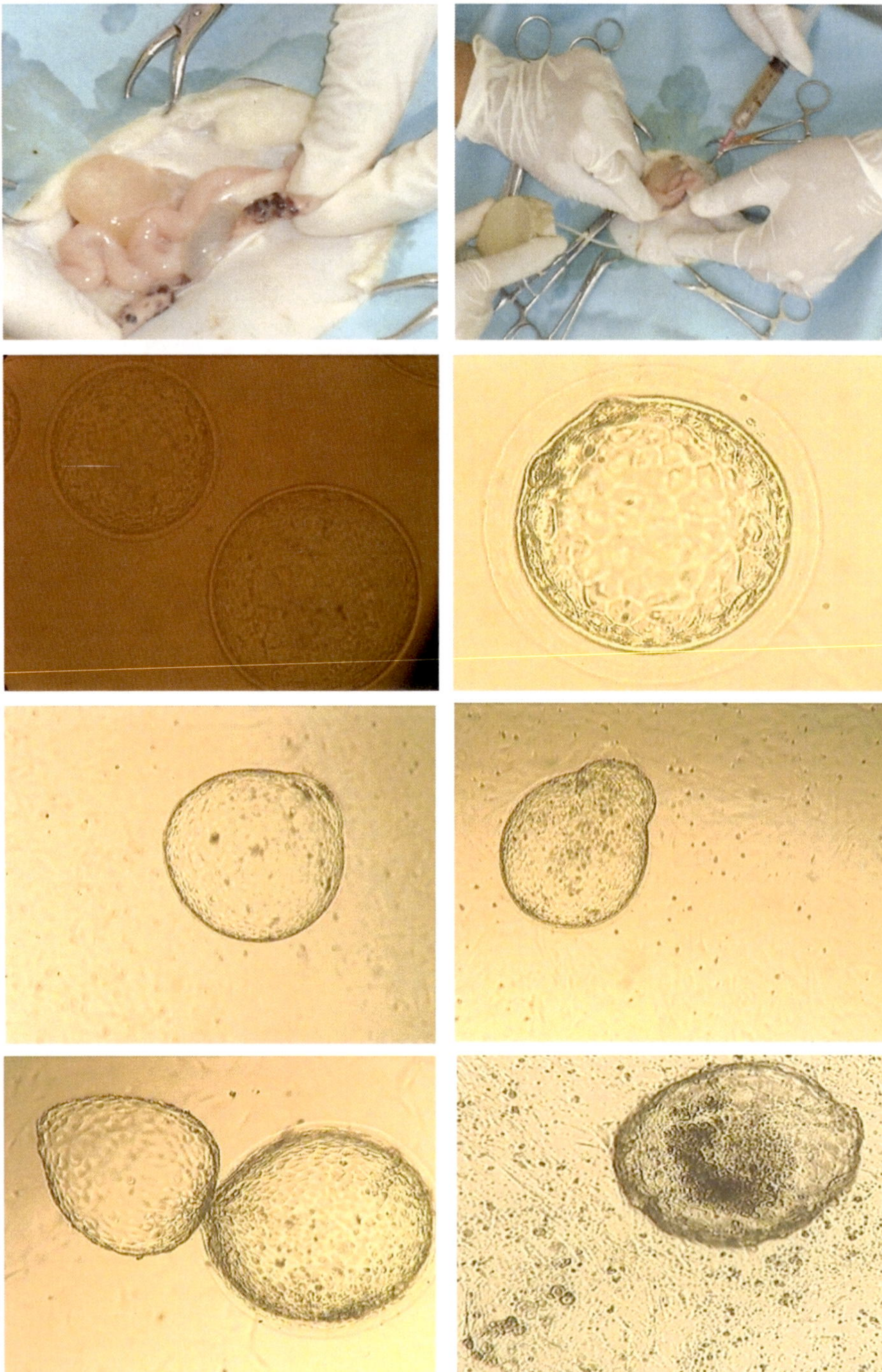

Procedure for embryo collection

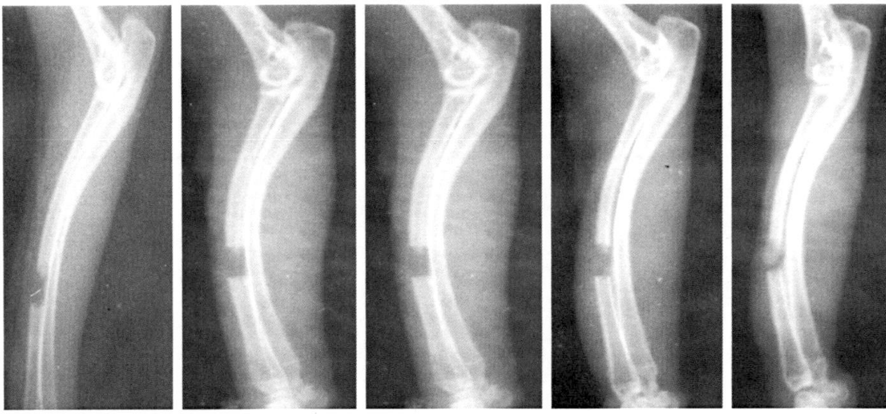

Day 0 10 20 30 40

Sequential radiographs of gr A

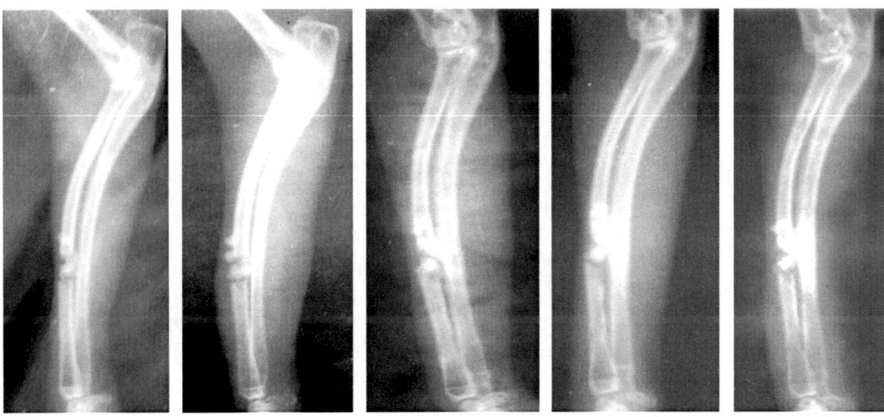

Day 0 10 20 30 40

Sequential radiographs of gr B

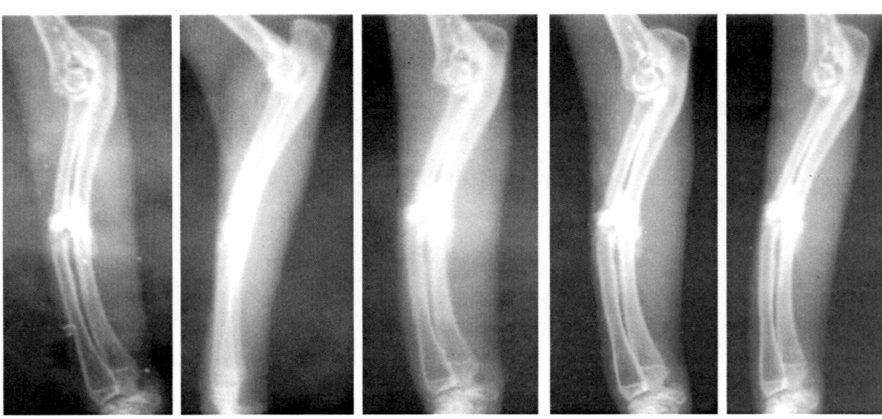

Day 0 10 20 30 40

Sequential radiographs of gr C

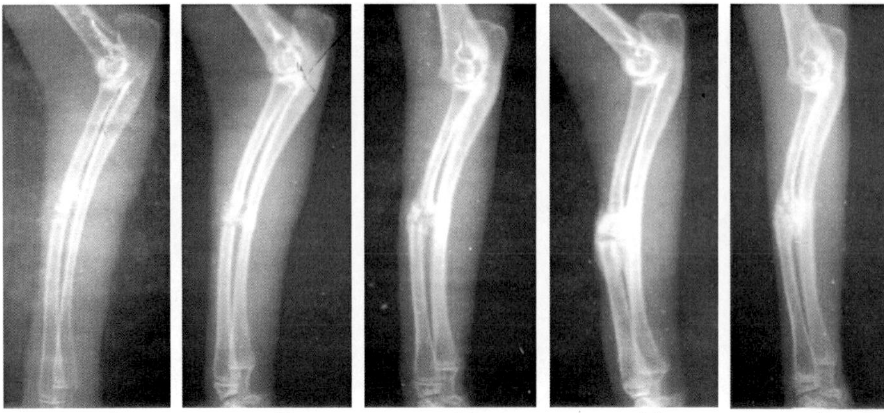

Day 0 10 20 30 40

Sequential radiographs of gr D

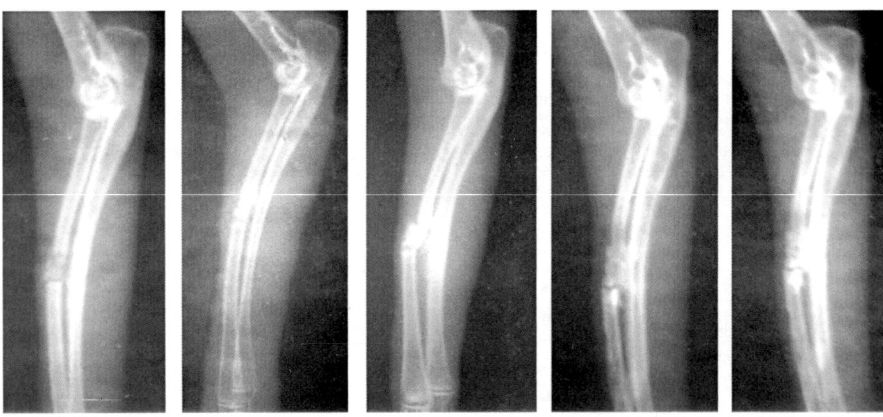

Day 0 10 20 30 40

Sequential radiographs of gr E

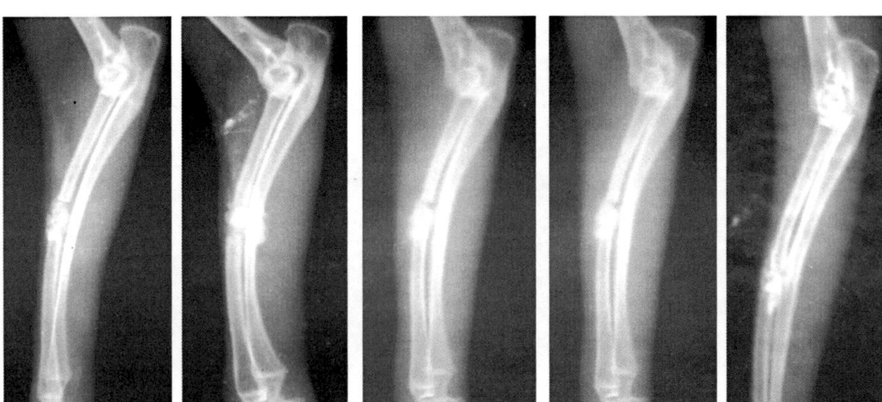

Day 0 10 20 30 40

Sequential radiographs of gr F

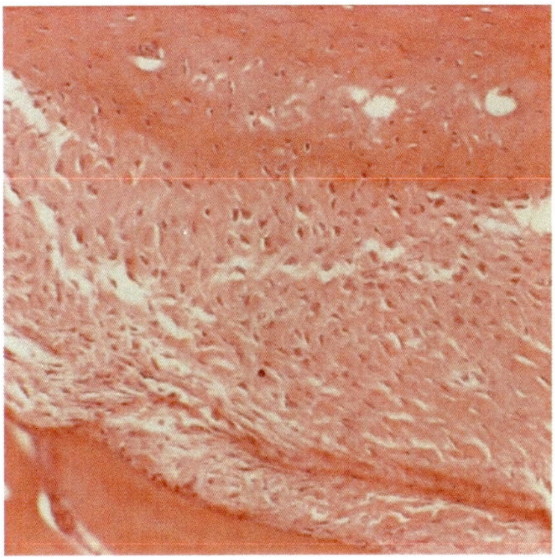

Group A

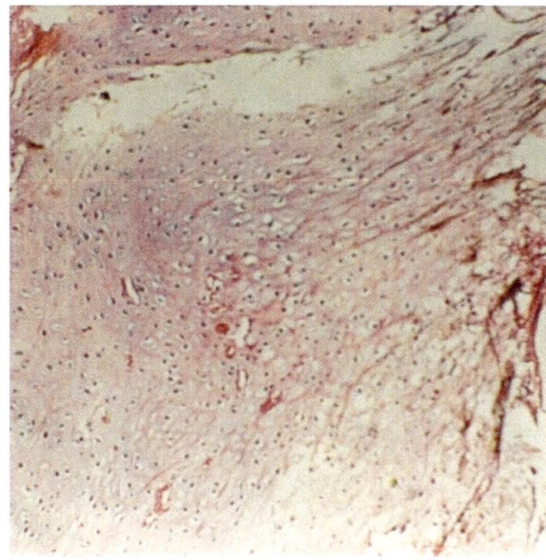

Group B

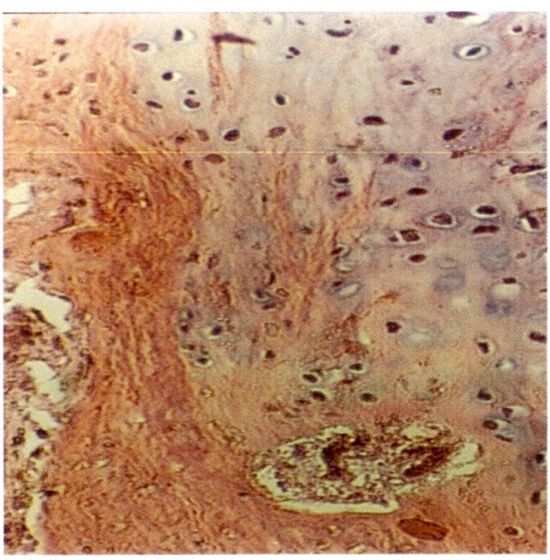

Group C

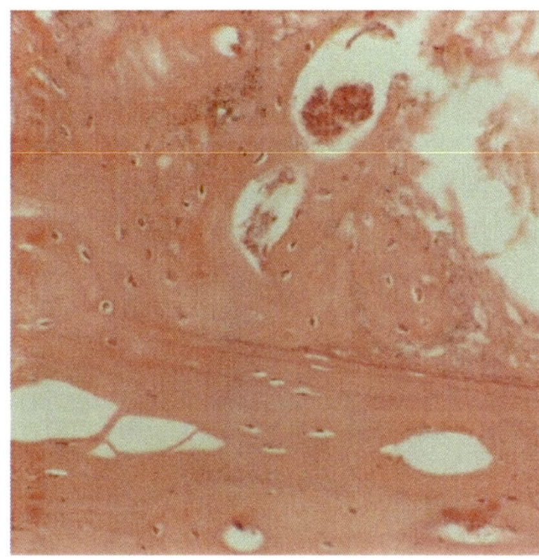

Group D

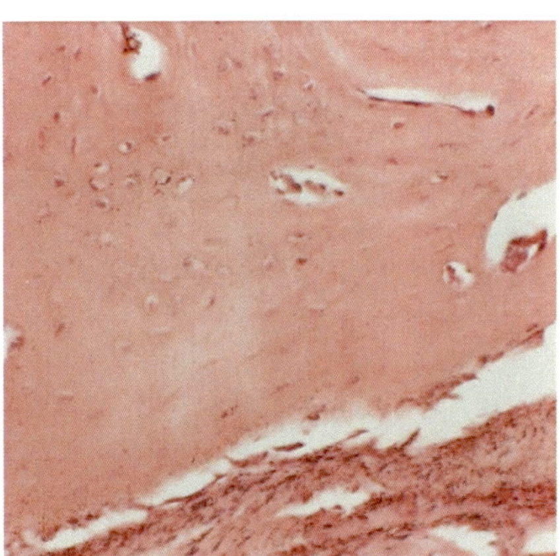

Group E

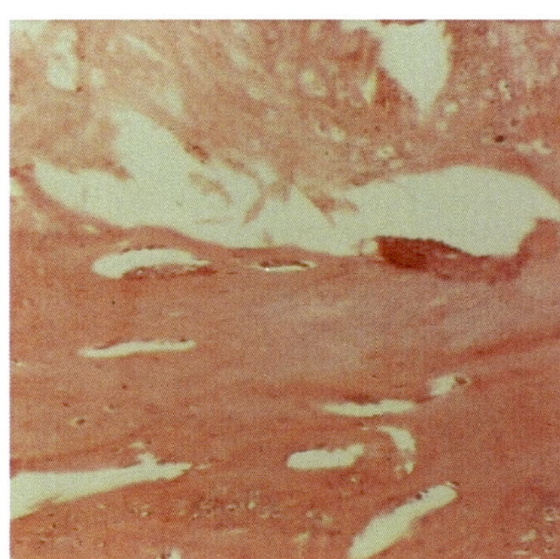

Group F

Histological observations of different groups

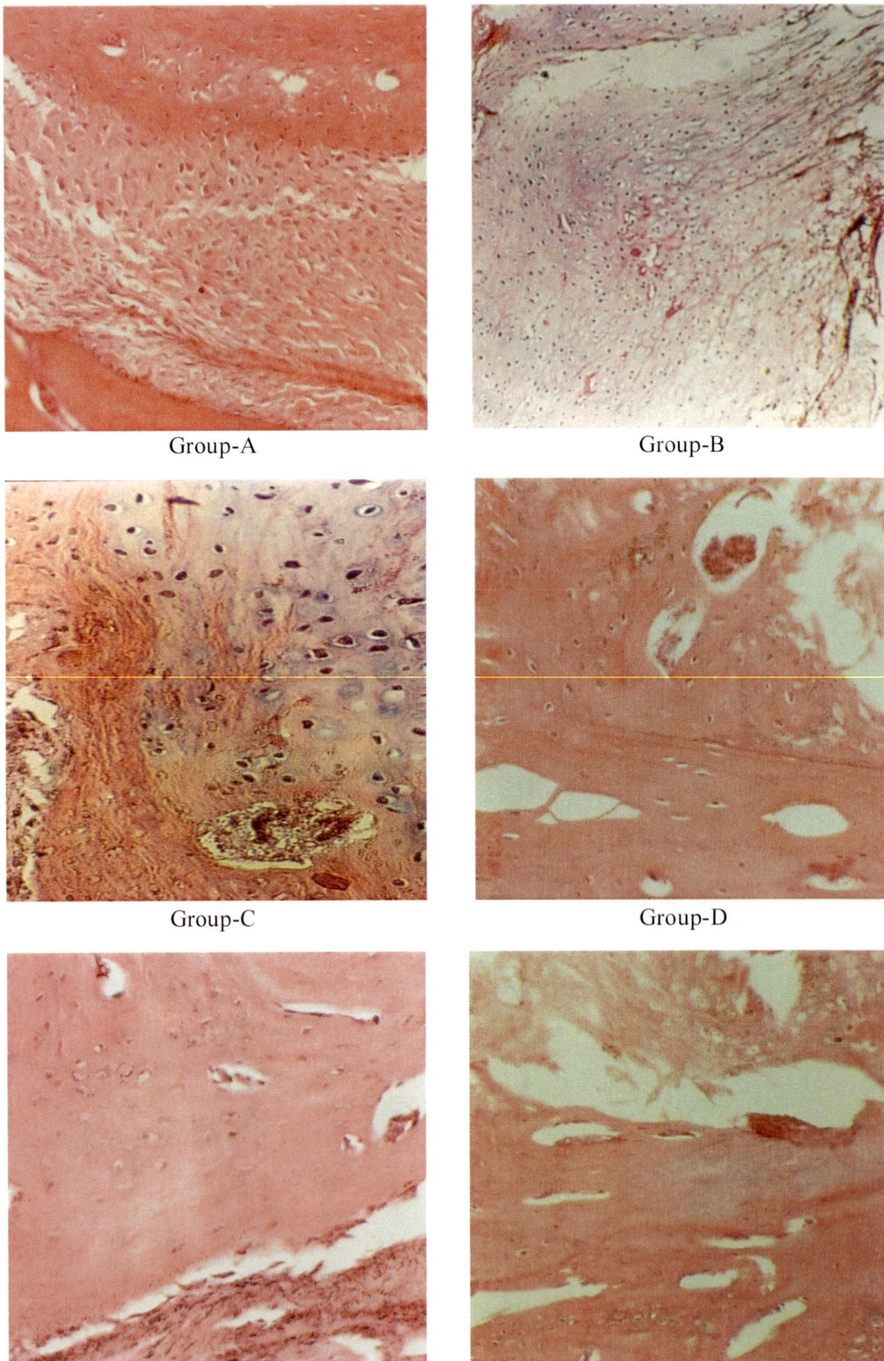

Group-A Group-B

Group-C Group-D

Group-E Group-F

Figure : Scanning electron microscopic (SEM)observations in different groups

13

Effect of Herbal Magacal on Fracture Healing

Fracture healing is a specialized type of wound healing response in which the regeneration of bone leads to restoration of skeletal integrity. Though the process of fracture healing is usually considered to be biologically optimum; the healing of 5-10% is delayed or impaired. For management of these types of fractures, use of biological methods for the enhancement of fracture healing is essential. Furthermore, fracture-healing can be modified by almost any endogenous or exogenous factor that has an influence on the metabolic function of cells. These factors include transforming growth factor-beta, bone morphogenetic proteins, hormones, vitamins, prostaglandin and calcium containing feed supplement medicine.

Keeping in view the above, in this chapter, efficacy of herbal product-Magacal in promoting osteogenesis during fracture healing in experimental animals is discussed. Magacal optimizes absorption, bioavailability and utilization of dietary calcium, phosphorus and magnesium. It is claimed to have advanced chelating agent to form organic complex chelate with dietary Ca, P and Mg, besides containing natural Vitamin D metabolites in highly bioavailability organic form, naturally protected and stable biotin and modern delivery system for optimum translocation of Ca, P and Mg ions to the sites of requirements.

Clinically healthy, adult New Zealand White rabbits of either sex (10) are divided equally into two groups A and B each consisting of 5 animals. All animals are kept on standard diet and management throughout the experiment. In all animals, midshaft simple transverse fracture is created in both the ulnae under thiopental (2.5%) anaesthesia. In test group (Group A), Magacal is given @5ml/animal/day, orally from day 5-20 post fracture, whereas, in group B, no drug is given which serve as control. Antibiotic dressing of the skin wound is done daily for one week with povidone iodine and himax ointment. All the animals are administered streptopenicillin @ 0.5 gm/animal/day for 5 days postoperatively. Skin sutures are removed on the10th postoperative day.

Clinically all the animals are observed for inflammation, pain and edema at the fracture site, surgical wound healing, appetite, general health, muscular atrophy of the limb, if any, at different intervals postoperatively.

Venous blood samples (5ml) from the ear vein are collected from all animals before and days 1, 3, 7, 15, 30 and 40 postoperatively. Out of this, 3 ml is used for serum collection and remaining 2 ml is collected in a vial containing anticoagulant EDTA. Freshly collected blood is analyzed for total leukocyte count (TLC), and differential leukocyte counts (DLC). The serum samples are analyzed for estimation of serum calcium, phosphorus, and serum alkaline phosphatase.

Lateral views of the X-ray at fracture site are taken just after surgery and on days 10, 20, 30 and 40 post-fracture to evaluate status of fracture reduction, progress of fracture healing and complications, if any.

The animals are sacrificed on day 40 and the fracture site is observed grossly for callus size and fate of surrounding soft tissues.

Bone sections are collected from the fracture sites involving the normal bone on either side after euthanizing the animals on day 40 post-fracture. After decalcification in Goodling and Stewart's solution (80% distilled water, 15% formic acids and 5% formalin) they are processed routinely for Haematoxylin and Eosin (H&E) staining to observe the status of fracture healing in two groups of animal.

For scanning electron microscopic study the bone specimens are collected from the healed bone immediately after euthanizing the animal on day 40. Every specimen is consisting of both fracture healing site and adjacent bone. They are fixed in 5% glutaraldehyde for 24-48 hrs. After fixing in glutaraldehyde, two washes for 30 min each with phosphate buffer saline (pH 7.4), followed by distilled water are performed. Then the specimens are dehydrated in serial concentration of ethanol from 30% to 100% for 30 min each. Thereafter, the specimens are dried with Hexamethyldisilazine (HDMS) for20 min. After that, golden ion sputtering is done on Jeol ion sputter (Model JFC 1100) at 7-10 mA and 1-2 KVP for 5 min. Finally, they are examined in scanning electron microscope (Jeol ISM 840 model) for the study of their different surfaces to identify the orientation and distribution of newly formed osseous tissue at the fracture site.

For undecalcified ground section study, oxytetracycline @25mg/kg b wt is injected intramuscularly on day 27, 28, 35 and 36, for double toning of the new bone. After euthanizing the animals on day 40, 2-3 mm thick sections are cut from the fracture site longitudinally including the surrounding normal bone. Undecalcified ground section are prepared and observed under ultra-violet light (Leitz Otrthoplan

Universal Wide field microscope with incidental light) to evaluate the nature and extent of the new bone growth at the fracture site.

All the animals of groups A and B maintained their normal appetite, defecation, and urination throughout the observation period. The general health of all the animals in both groups is also good throughout the experiment. Inflammation at the fracture site increased upto fourth postoperative days in all animals of both groups, the test limb is then found normal in appearance by day 7. Surgical wound healing is by first intention in all the animals. Earlier return of weight bearing is seen in animals of test group followed by control.

Blood cell count shows significant increase in total leukocyte count (TLC) in all the animals from the very first postoperative day. On subsequent intervals, there is no marked difference when compared between two groups at different intervals. Differential leukocyte count (DLC) show a significant increase in neutrophils and concomitant decrease in lymphocytes upto day 7 post-fracture, whereas absolute increase in neutrophil is noticed till day 15 and monocytosis is noticed on day 30 in animals of group B (control) which then gradually decreased towards normal values on day 40. In animals of group A, where herbal Magacal is given, initial neutrophilia is noticed upto day7 and it gradually returned to normal values at subsequent intervals. The other cells like basophilic and eosinophilic counts fluctuated within normal range at different intervals.

The serum calcium levels in animals of test group shows non-significant increase upto day 7 and then significant decrease on days 15 and 30, which finally reached towards normal level by day 40 gradually. In the animals of group B (control), it increase non-significantly upto day 15 and then it decrease progressively till the end of observation. When compared between the two groups, serum calcium levels are significantly low in test group on days 15 and 30 as compare to control group, as more amount of bridging callus is formed in the test group of animals. Calcium of bone callus is stated to be derived from the serum calcium as seen in the test group as compare to control group, where bridging callus formation is relatively small and delayed.

Serum phosphorus level increased immediately after fracture fixation in animals of both groups and reach maximum level on third postoperative day. In subsequent intervals, it progressively improves with some fluctuations and remains slightly higher than the base value in all groups at the end of observation period. The increase in serum phosphorus level during early period of fracture healing may be due to necrotic disintegration of cells at the site of fracture.

Serum alkaline phosphatase (AP) actively increase significantly from day 3 onwards and reached its maximum on day 15 and then decrease to normal by day 40 in the animals of Magacal administer test group, whereas, AP activity is maximum on day 40 in the control group of animals. Alkaline phosphatase is one of enzyme which is abundantly present in the ossifying zone of a long bone which hydrolyses the organic esters resulting into increased organic phosphate for bone formation. The increased AP activity in the early post fracture period in the Magacal administered test group may be suggestive of an increase osteoblastic proliferation during fracture repair.

Radiographs taken immediately after surgery in test groups (A) shows radiolucent diaphyseal transverse defect in the ulna, whereas, radius acted as natural splint. On day 10, there is periosteal reaction extending to almost entire diaphysis of the ulna and both the fracture ends become irregular and hazy. This periosteal reaction appears more pronounce on day 20. At this stage, the fracture gap is almost filled by newly formed osseous tissue and bridged the bony defect. On day 30, there is marked reduction in periosteal reaction as bridging callus is organized and osseous union is completed. On day 40, there is marked reduction in callus size and smooth. Uniformly dense periosteal surface clearly indicates remodelling of healing site. Periosteal continuity across the fracture line is completed. The union of medullary cavity across the fracture line, however, is not completed.

In control group (B) radiographs of days 0, 10 and 20 shows that fracture ends though become rough and uniform but periosteal as well as endosteal proliferation along the defect is relatively slow compare to test group (A). Radiographs of day 30 reveals new bone formation from the cut ends as well as adjacent periosteal surface of radius, but fracture gap is well visible. At day 40, these periosteal proliferations are more intense. The defect size reduces but still it is well visible at day 40, suggesting delayed healing compare to test group. At this stage, periosteal proliferation have bridged with adjacent radius, which acts as a natural splint throughout the experimental period.

Histopathologically, in the Magacal administer test group (A), complete union at the fracture line with excellent remodelling is observed. The fracture gap is filled with newly developed bony trabeculae, which is evinced by its direction and is more woven in nature. Some newly formed thickened bony trabeculae penetrating from both the fracture ends are also noticed. The periosteum is also thickened and infiltrated with mononuclear cells. Resorption and replacement is under progress, at both ends. In group B (control), incomplete union at the fracture site is evident and the fracture gap and newly formed bone are well distinguished from that of normal bone. The newly formed osseous tissue is evinced in an improper direction and is not fully developed and does not crossover the fracture line. Blood vessels and mononuclear infiltration are also evident within the marrow tissue.

In the animals of test group (A) where Magacal is administered as an adjunct, excellent fracture healing is observed as evident by complete bony union at the fracture fragment and restoration of continuity of cortex. Under fluorescent microscope, the fracture line is visualized as golden yellow fluorescence in between the fragments, whereas, the host bone evinced dark sea green homogenous colour. Within the new osteoid tissue, which completely filled the fracture gap, crossing over of numerous new bone trabeculae is evident in oxytetracycline labeling study. Resorption cavities are present on either ends, indicating that the resorption and replacement of the bone is well under progress. The resorption cavities are indication of initiation of bone remodelling. In the Magacal administered test group, ground sections shows good fracture healing with new bone having density similar to old bone. This suggests that the healing is comparatively faster in these animals and the bone regained new normal appearance.

In the animals of control group (B) incomplete union at the fracture site is evident by the presence of newly formed less dense osteoid tissue and the continuity of the cortex across the fracture line is also not restored. Under UV light, relatively less dense fluorescence is seen at places suggesting that the fracture gap is not completely filled with new osteoid tissue. A gap is seen at the cranial edge filled with fluorescent new bone suggesting new bone formation which fails to cross over to other end, which results in delayed union.

The scanning electron microscopic (SEM) observations provide the status of fracture healing in both the groups. It also provides topographic picture about the type of tissues involves in the healing process and status of vascularity at the fracture site. In animals of Magacal treated group (A), SEM clearly indicates osseous union at both the fracture ends, however, process of remodelling is still continuing. This is evidenced by the presence of bone cells and un-mineralized collagen network at the defect site. Further magnification of the section shows that the healing site appears hyper vascularized and new vessels penetrates into the healing site. The origin of such vessels is mostly from the periosteum and surrounding soft tissues. In animals of group B (control), the SEM picture shows that the bone defect is not completely healed but partially filled with fibrocartilagenous tissue. The osseous tissue is predominantly seen along the fracture ends, whereas, the fibro cartilage is present at the center of the defect. The process of lying down of bone matrix and vascularisation is also comparatively less in this group of animals.

Feeding of herbal Magacal to the animals of test group might have activates/ stimulate the osteoprogenitor cells in the body which in turn results in early proliferation of the osteogenic cells and formation of organic matrix. Radiological, histological, scanning and undecalcified ground section study confirms that the Magacal administer test group of animals osteogenesis starts much earlier than in control group of animals. The herbal Magacal probably creates some cellular environment as osteoinducers which causes early osteogenesis in the test group of animals. Different evaluation parameters also suggests that not only early osteogenesis, but also early vascularisation and remodelling takes places in the Magacal administered test group of animals compare to control animals.

In conclusion, herbal Magacal accelerates fracture healing in a controlled experimental study.

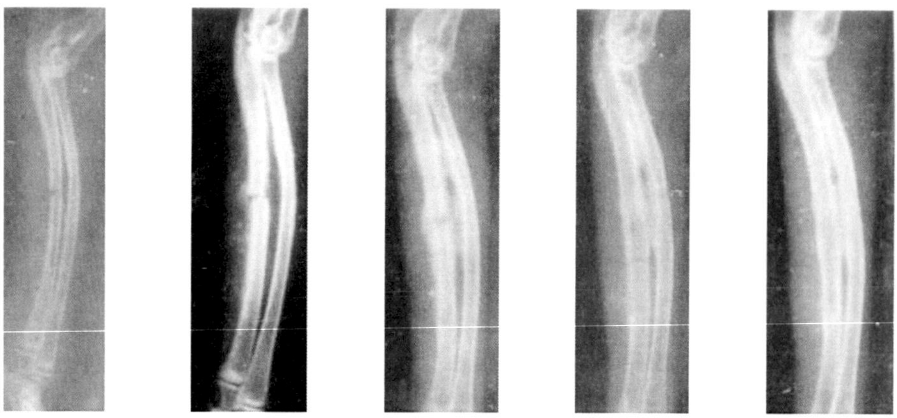

Day 0 10 20 30 40

Group A

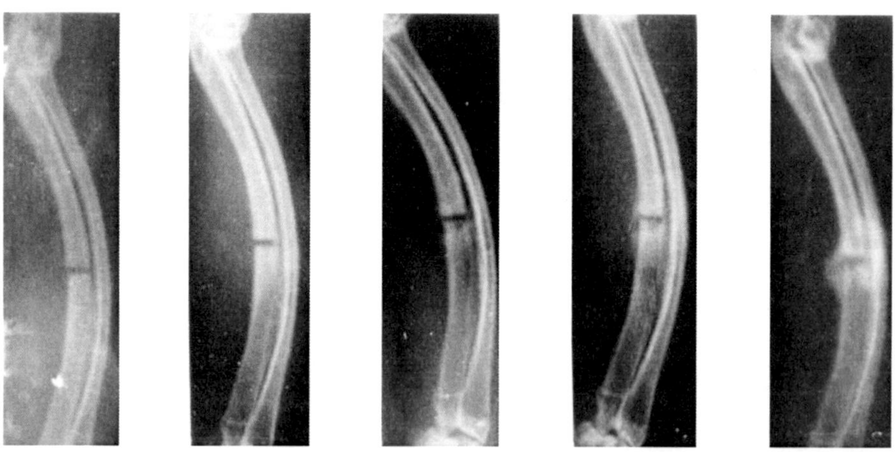

Day 0 10 20 30 40

Group B

Lateral radiographs of Magacal fed test group (A) showing accelerated course of fracture healing, whereas, in control group (B) showing relatively delayed fracture healing in subsequent same intervals.

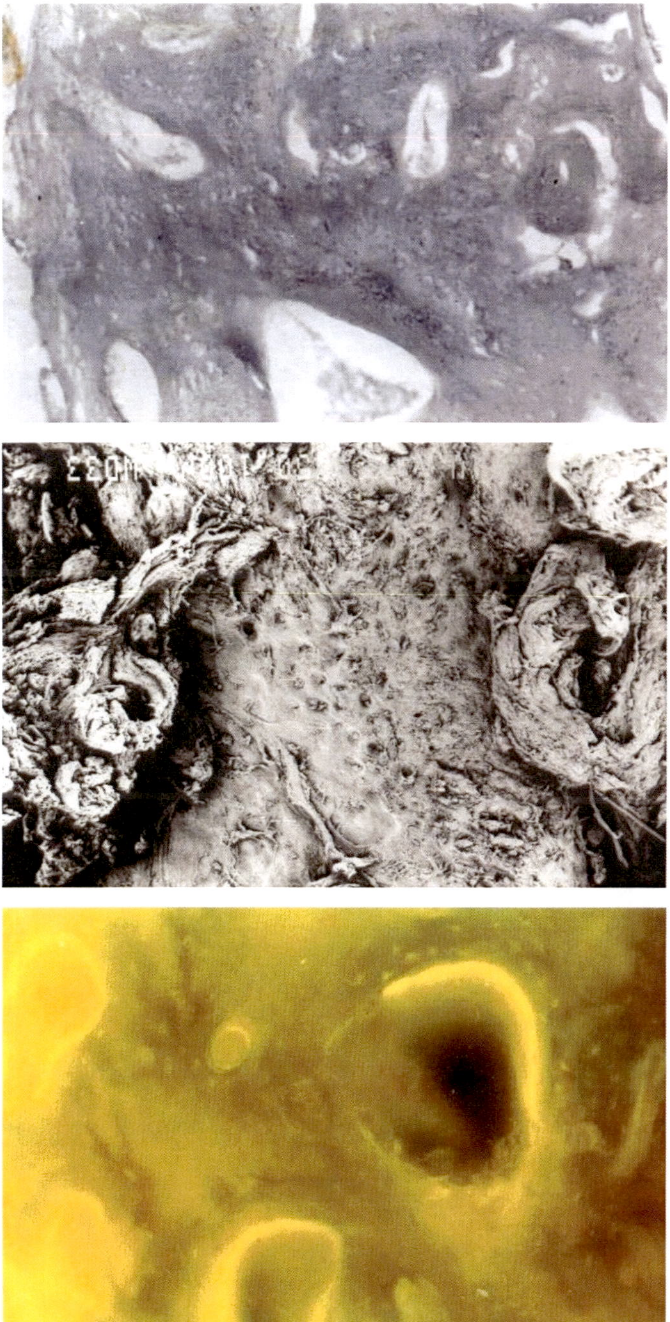

Photomicrographs of H&E, SEM and OTC labeling studies shows abundant newly formed bony trabeculae penetrating from both the fracture ends and healed completely at day 40 post fracture in the Magacal administered test group (A).

14

Effect of Bone Wax on Fracture Healing

During orthopaedic surgery, it is very difficult and sometimes impossible to stop the flow of blood from fractured bone. For this reason, there has been a long-standing interest in haemostatic agent for osseous tissue surgery. Bone –wax, oxidized cellulose, sanguistat, bone dust, marrow paste and combination using polyethylene glycols have been tried for osseous hemostasis, however, bone wax appears to be the most promising bone haemostatic agent and is widely used today. But, very littlie work has been done to define the local effect of bone wax on healing and osteogenesis and its eventual fate in animals.

The objective of this study is thus, undertaken in this chapter to investigate the effects of bone wax on cortical bone healing.

Clinically healthy adult New Zealand white rabbits (10) of either sex are divided equally in to two groups A and B, consisting of 5 animals each. All animals are kept in standard diet and management throughout the experiment. In all animals' mid-shaft simple transverse fracture are created in both the ulnae under Xylazine-Ketamine anaesthesia. In animals of group A, bone wax packed in between two fracture ends, whereas, in animals of group B, fracture site kept as it is.

Lateral and anterio-posterior radiographs are taken on days 0, 10, 20, 30 and 40 post-fracture. The radiographs are observed for progress of fracture healing and complication, if any at different interval. The specimens for histopathological study are collected from the healed site of test bone immediately after euthanizing the animals on day 40. Every specimen is consisted of both healing site and adjacent host bone. The bone sections are preserved in 10% formalin for 9-10 days and then are decalcified using Goodling and Stewart's solution (15% formic acid, 5% formalin and 80% distilled water). Completely decalcified bone sections are then processed routinely for H&E staining. These sections are observed for the status of healing at the fracture site at cellular level.

Inflammation at the fracture site is increased up to day 4 postoperatively in all the animals of both groups, and then the limb is normal in appearance by day 7. Surgical wound healing is by first intention in all animals. Earlier return of weight bearing seen in animals of control group followed by test. In test group of animal (A) radiography taken immediately after surgery shows radiolucent diaphyseal transverse defect in the ulna, whereas radius acted as a natural splint. On day 0, the fracture fragments are perfectly aligned and bone wax is not visible. Subsequent radiography of days 10, 20 and 30 shows fracture gap gradually increases, fracture ends became smooth and periosteal as well as endosteal new bone proliferation along the defect site is relatively slow as compared with the control group in same interval. However, day 30 radiography reveals very little periosteal reaction started from proximal end of fracture towards adjacent radius. This reaction is pronounced on day 40. In this stage, periosteum became thickened and periosteal proliferation bridging with the adjacent radius but failed to obliterate the fracture gap completely. Due to this periosteal reaction, fracture site became hazy and cloudy, but bony union is not taking place. No sign of any infection is observed in any animal at any interval.

In group B, periosteal reaction is started as early on day 20. On day 30, this periosteal reaction is extended to almost entire diaphysis of the ulna and also in the adjacent radius and the fracture gap is reduced. On day 40, bridging callus became more intense and organized and it is almost obliterate the fracture gap. In this stage, periosteal callus is also bridged with the adjacent radius adjacent to the fracture site.

Histopathologically, in the bone wax treated group (A), incomplete union at the fracture site is evident, and the fracture gap and newly formed very little osseous tissue is well distinguish from that of normal bone. In some animal of this group, bone wax is still visible in the center and there is very little evidence of osteogenesis as the margin or even osteogenesis is completely inhibited, leaving a cystic defect. The newly formed osseous tissue, which has developed very little, is evinced in an improper direction and is neither fully developed and nor crossed over the fracture lines. The inflammatory reaction is minimal, and, although few giant cells are seen at the periphery of the wax, this reaction is limited.

In group B (control), the fracture gap is almost filled with newly developed bony trabeculae which is evinced by its proper direction and is woven /immature in nature. In some animals of this group, there is some evidence of reorganization of this immature bone and development of new marrow representing the total reparative process. Some newly formed thickened bony trabeculae penetrating from both the fracture ends are also noticed. The periosteum is also thickened and infiltrated with some mononuclear cells.

This haemostatic material for bone is a combination of bees wax, almond oil and salicylic acid and is commonly known as bone wax. Its haemostatic action is physical (tamponade) rather than chemical. Its assets include ready availability, ease in preparing and handling and quick results.

From the radiological and histopathological observations, it is revealed that the presence of bone wax at the fracture site almost completely inhibited the formation of new bone. This effect appears to be related to the impenetrable, non-resorbable nature of wax. The wax apparently represented as inert, physical barrier to the reparative process. Also of considerable interest is the remarkable lack of foreign body reaction. The eventual fate of bone wax is still open to question, as local macrophages shows essentiality no evidence of mobilizing the material even as late as 40 days. It is also reported that the effect of bone wax is localized, as completely inhibiting osteogenesis. One possible indication for the use of bone wax is for treatment of recurring exostosis or osteophyte near articulating surface, but there seems to be no universal knowledge or acceptance of the idea.

Summary

The effect of bone wax on fracture healing is studied in rabbits as an experimental study. Bone wax is found causes delayed fracture healing or even inhibit the formation of new bone at the fracture site ultimately leads to non-union.

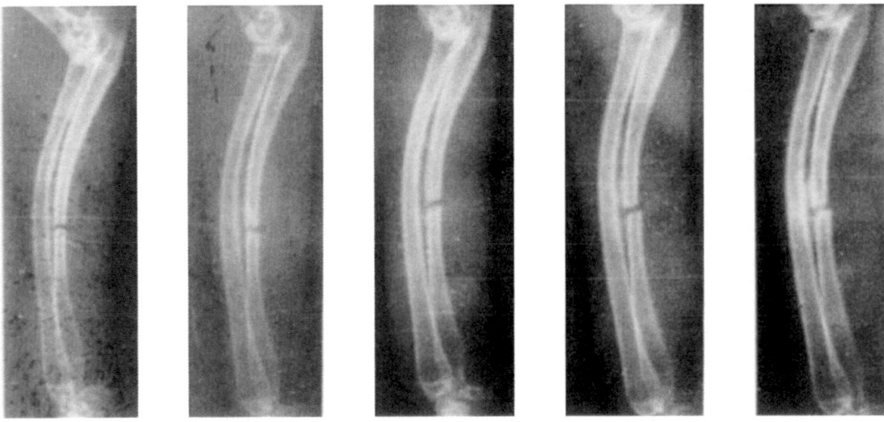

Day 0 10 20 30 40

Group A

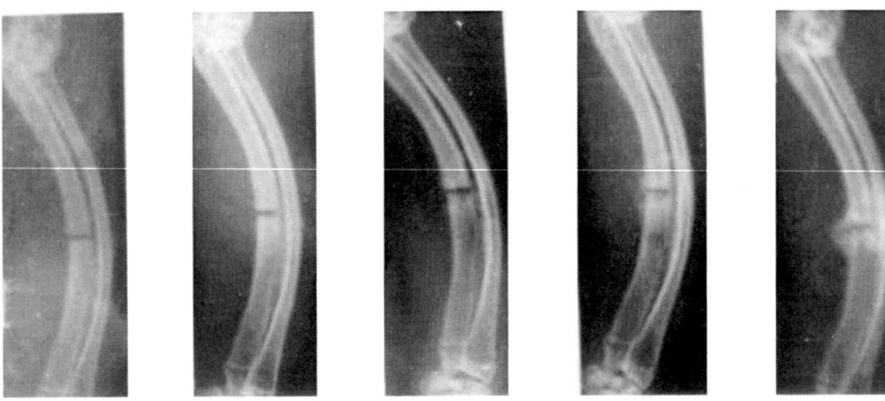

Day 0 10 20 30 40

Group B

Lateral radiographs of test group (A) showing delayed course of fracture healing; radiographs of group B showing normal course of fracture healing

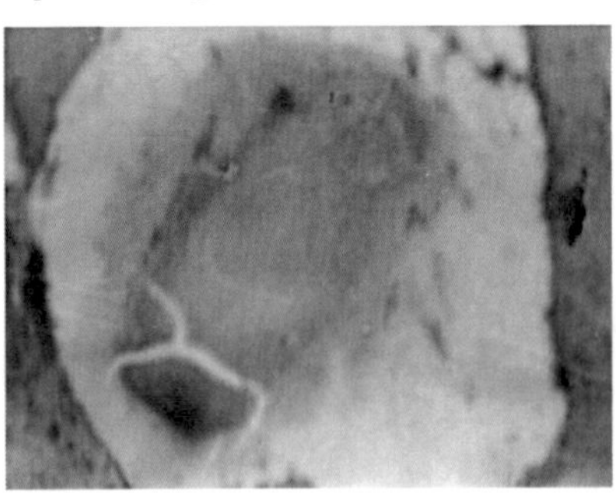

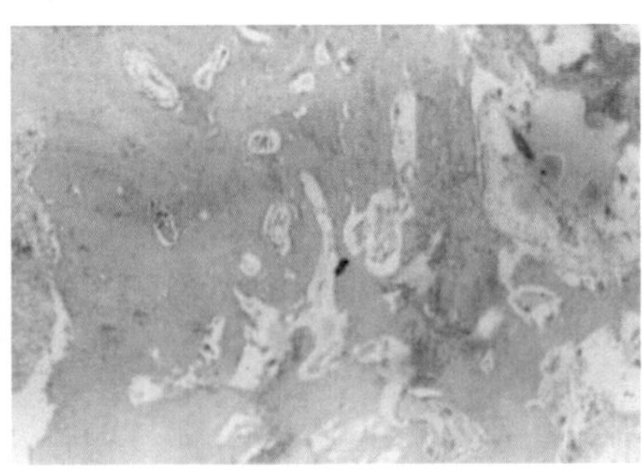

Histopathological photomicrograph of group A (left) showing bone wax still visible in the center and less evidence of osteogenesis; group B (right) showing new immature bony trabeculae at the fracture site.

15

Effect of Prostaglandin (PGE_2) on Fracture Healing

Bone healing is an extraordinary process involving not only an inflammatory reaction and formation of callus but also reconstruction of the defect to regain its original biomechanical strength. Bone healing can be modified by variety of endogenous or exogenous factors that affect the metabolic function of bone cells. Prostaglandins (PG), a family of cycled derivation of unsaturated long chain fatty acids, are intimately associated with bone metabolism, and are important as locally active agents with significant effects on bone and bone cells *in vitro*. In some experimental studies, it is suggested that the local increase of prostaglandin in response to trauma may act to stimulate differentiation and proliferation of osteoprogenitor cells during fracture healing, whereas, specific prostaglandin antagonists, indomethacin, inhibited fracture healing. The objective of this chapter is to investigate the effects of locally (defect-site) administered prostaglandin (PGE_2) on cortical bone healing in rabbit model.

Clinically healthy adult New Zealand White rabbits (10) of either sex are divided equally into groups A and B, consisting of 5 animals each. They are kept on standard diet and management throughout the experiment. In all the animals, mid-shaft single transverse fracture is created in both the ulnae under thiopental (2.5%) anaesthesia.

In test group A, 0.1 ml (25 μg) prostaglandin (PGE_2) is injected at fracture site on alternate day for 30 post operative days, whereas, group B is served as control. In both the groups lateral and antero-posterior radiographs are taken on days 0, 10, 20, 30 and 40 postoperatively to observe progress of fracture healing and complication, if any at different intervals.

The specimens for histopathological studies are collected from the healed site of test bone immediately after euthanizing the animals on day 40. Every specimen is consisted of fracture healing site with adjacent host bone. The bone sections are preserved in 10% formalin for 9 to 10 days and then they are decalcified using Goodling and Stewart's solution (15% formic acid, 5% formalin and 80% distilled water). Completely decalcified bone sections are then processed routinely for H & E staining. These sections are observed for status of healing at fracture site.

Inflammation at the fracture site is increased up to fourth postoperative day in all the animals of group B and then the limb is normal in appearances by day 7, whereas relatively prolonged inflammatory phase (up to day 10) is observed in group A, probably due to exogenous administration of prostaglandin in this group of animals. However, surgical wound healing is by first intention in all the animals. Earliest return of weight bearing is seen in animals of test group followed by control group. No physiological side effects and other complications are observed in any group.

In animals of group A, radiographs taken immediately after surgery shows radiolucent diaphyseal transverse defect in the ulna whereas radius acted as a natural splint. Both the fractured ends are rough. Radiographs taken at subsequent intervals (days 10 and 20) indicates periosteal proliferation from both the cut ends as early as on day 10 postoperatively and defect site is reduced and appeared elliptical in shape. At day 20, the periosteal proliferation is more intense and further decrease in the gap of bony defect is seen. Radiographs of day 30 of group A shows that the bone defect is completely filled with newly formed osseous tissue i.e. periosteal, endosteal and inter-cortical callus. The periosteal new bone is well organized and is in the process of remodeling. The medullary cavity across the bony defect looks completed. Day 40 radiographs further reveal periosteal and medullary continuity at the defect site and remodeling of bridging callus are well formed. In animals of control group B, radiographs of days 0, 10 and 20 shows fracture ends though became rough and uniform but periosteal and endosteal proliferation along the defect is relatively slow as compared with the test group (A). Radiographs of day 30, revealed new bone formation from both the cut ends and adjacent periosteal surface of radius, but fracture gap is well visible. At day 40 these periosteal proliferation are more intense. Though the defect gap is reduced in size, but it is well visible at day 40, suggesting delayed healing as compared to test group.

Histopathologically, complete union and excellent remodeling at the fracture site are observed in the animals of test group A. Newly formed osteoid tissue is seen lying adjacent to the old bone and in all the animals of this group, the fracture line is indistinct. Abundant bony spicules are present at the fracture site with the presence of osteoclast around them. At places, newly laid down trabecular bone is seen crossing over on either ends at the periosteal and endosteal sides and is consolidated and thickened. The periosteum is also thickened and infiltrated with mononuclear cells. There is abundance of marrow cells with presence of osteoclastic activity. Resorption cavities and replacement by newly formed osseous tissue is the characteristic feature of this group. In (control) group B animals, incomplete union at the fracture site is evident and the fracture gap and newly formed bone are well distinguished from that of normal bone. The newly formed osseous tissue near normal bony tissue is evinced in an improper direction and is not fully developed and did not cross over the fracture line. Blood vessels and mononuclear infiltration is also evident within the marrow tissue.

Prostaglandin (PGE_2) is known to stimulate cyclic amino phosphate production by isolated osteoblast cells and produce contractile shape changes that suggest activation of such cells. When injected locally *in vivo,* PGE_2 produces a variable effect on resorption and increased formation in calvarial bone. The systemic effects of PG have been less well studied. Long-term intravenous therapy with PGE_2 to maintain circulation in infants with cardiac defects leads to periosteal proliferation on the surface of long bones. PGE_2 given orally for 90 days produces a dose response increase in periosteal bone formation on femoral cortical of dogs. This is associated with increase activation of remodeling activity within the cortical bone itself, which is also found in this study. It's reported that in rats, the amount of trabecular bone in the tibial metaphysis is increased in a dose response fashion by administration of PGE_2 for 3 weeks. Much of this appeared to be new woven bone formed in the marrow tissue that suggested *de novo* activation of bone formation and associated with increased number of osteoprogenitor cells and osteoblast. It is reported that when prostaglandin (PGE,) injected locally in dogs resulting in more periosteal and cortical endosteal bone formation, and increased resorption surface in the cortical endosteal envelope, indicating accelerated remodeling. In contrast, of the above fact, higher concentration (> 10 -6 M) of prostaglandin suppressed the cellular activities of cultural bone cells.

Conclusion

PGE_2 in the smaller dose schedule has a local effect on long bone fracture healing and remodeling that involves stimulation of bone formation activity *in vivo.*

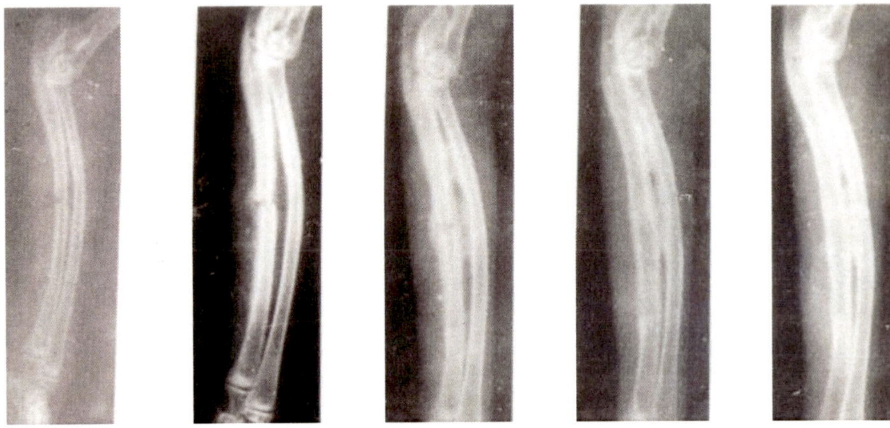

Day 0 10 20 30 40

Group A

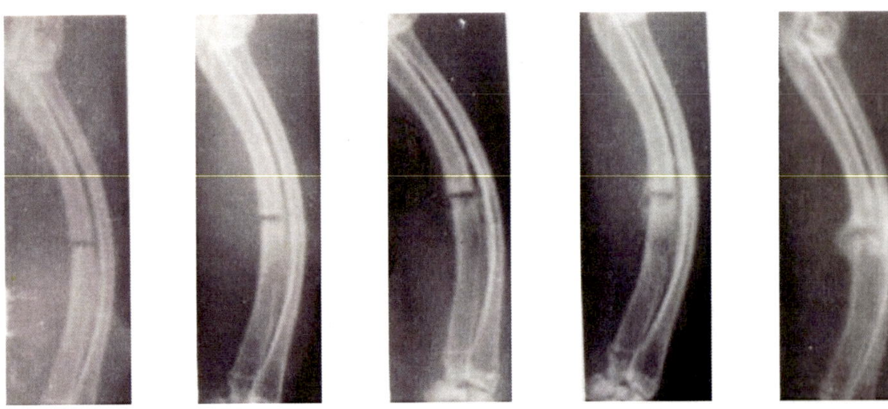

Day 0 10 20 30 40

Group B

Lateral radiograph of test group A showing accelerated fracture healing, group B (control) showing relatively delayed fracture healing in different intervals.

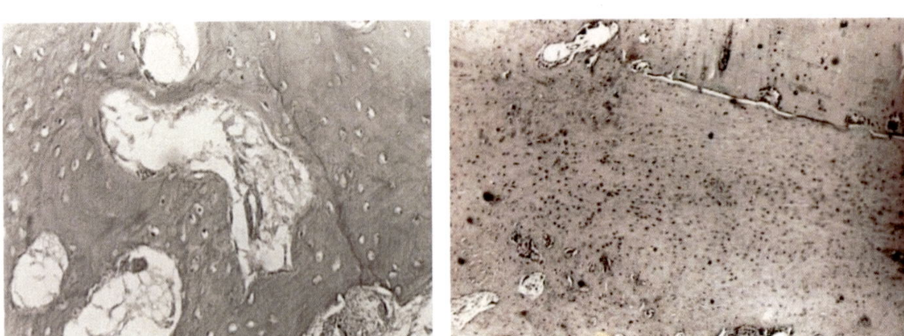

Histopathological photomicrograph of test group (left) showing abundant new bony trabeculae crossing over the fracture site, group B showing well visible fracture line

16

Effect of Gonadotropic Hormone (Gnrh) on Fracture Healing

The action of the gonadial hormone is not confined to the growth and development of the sexual organs alone, but the normal body growth as well as the growth of the skeleton are to a great extent influenced by these hormones. Of the female gonadial hormones, the action of estrogen and progesterone on skeletal growth has not been studied in detail. Some reports implicate estrogen deficiency in post menopausal osteoporosis and shows that estrogen replacement therapy can help to prevent that post menopausal stage of accelerated rate of bone loss. It's reported that the progesterone treatment play a significant role in regulating cortical bone cell activity and as such, suggest that progesterone may play a part in the prevention and slowing down of cortical bone loss in females following menopause or oophorectomy. The effect of gonadotropic hormone on bone healing has not yet been studied in detail. The chapter is thus, undertaken to investigate the effects of gonadotropic hormone (Gn-rh)) on cortical bone healing in female rabbits.

Clinically healthy adult female New Zealand white rabbits (10) are divided equally into groups A and B each consisting of 5 animals. All animals are kept on standard diet and management throughout the experiment. In all animals midshafts simple transverse fracture is created in both the ulnae under thiopental (2.5%) anesthesia. In test group (group A) 0.1 ml of Gn-rh analogue is injected intramuscularly at 3 days interval up to 30 days post operatively.

Lateral and anterio-posterior radiographs are taken on days 0, 10, 20. 30. and 40-post fracture. These radiographs are observed for progress of fracture healing and complication, if any, at different intervals. The specimens for histopathological study are collected from the healed site of test bone immediately after euthanizing the animals on day 40. Every specimen consists of better healing site and adjacent host bone. The bone sections are preserved in 10% formalin for 9 to 10 days and then are decalcified using Goodling and Stewart's solution (15% formic acid, 5% formalin and 80% distilled water). Completely decalcified bone sections are then processed routinely for H & E staining. These sections are observed for status of healing at fracture site.

Inflammation at the fracture site increases up to fourth post operative day in all the animals of both groups, then the limb is normal in appearance by day 7. Surgical wound healing is by first intention in all the animals. Earlier return of weight bearing is seen in animals of test group followed by control.

In test group of animal (group A) radiographs taken immediately after surgery shows radiolucent diaphyseal transverse defect in the ulna, whereas, radius acted as a natural splint. On day 0, the fracture fragments are perfectly aligned. On day 10, there is periosteal reaction is extending to almost entire diaphysis of the ulna and also in the adjacent radius. This periosteal reaction appears more pronounced on day 20. At this stage, the fracture gap is almost filled by newly formed osseous tissue and bridged the defect. On day 30, there is marked reduction in periosteal reaction as bridging callus is organized and osseous union is completed. The osseous union between radius and ulna adjacent to fracture site is also appreciable. Periosteal continuity is well advanced. On day 40, marked reduction in callus site and smooth, uniformly dense periosteal surface can appreciate remodeling of healing site. Periosteal continuity is well developed at the fracture site. The union of medullary cavity across the fracture line, however, is not complete. In the control group (B) subsequent radiographs of days 0, 10 and 20 shows that fracture ends though become rough and uniform but periosteal as well as endosteal proliferation along the defect is relatively slow compare to the test group (A). Radiographs of day 30, reveals new bone formation from both the cut ends as well as adjacent periosteal surface of radius, but fracture gap is well visible. At day 40, these periosteal proliferations are more intense. The defect size is reduced but it is well visible at day 40 P0 suggesting delayed healing compare to test group. In this stage, periosteal proliferation is bridged with the adjacent radius.

Histopathologically, in the test group (A), complete union at the fracture line with excellent remodelling is observed. The fracture gap is filled with newly developed bony trabeculae, which is evinced by its improper direction and is more woven in nature. Some newly formed thickened bony trabeculae penetrating from both the fracture ends are also noticed. The periosteum is also thickened and infiltrated with mononuclear cells. Resorption and replacement is under progress, at both the ends. In group B (control), incomplete union at the fracture site is evident and the fracture gap and newly formed bone is well distinguished from that of normal bone. The newly formed osseous tissue is evinced in an improper direction and is not fully developed and does not crossover the fracture line. Blood vessels and mononuclear infiltration is also evident within the marrow tissue.

It is reported that progesterone and estrogen supplementation to the spayed dam resulted in an enhancement of individual bone cell activity. Progesterone and estrogen supplementation altered cortical/ trabecular bone remodeling activity in the absence of ovarian hormones. I also reported that progesterone supplement in spayed rabbit accelerate cortical bone healing. In the present study, gonadotropic hormone in these dose schedules accelerates cortical bone healing. The exact mechanism of action of this hormone in bone healing is still unknown, but it increases production of sex hormone i.e. estrogen and progesterone which, in turn, accelerated bone healing. One would have to consider the possibility of competitive bindings of sex hormone for glucocorticoids receptors on bone cells.

Summary

The effects of gonadotropic hormone (Gn-rh) analogue on fracture healing are studied in rabbits experimentally. The fracture healing is evaluated based on radiological and histopathological examinations. In test group, periosteal proliferation and bridging callus is developed by day 20 and complete obliteration of fracture gap is observed on day 30 of post-fracture. Remodelling of bridging callus and well developed periosteal continuity are observed on day 40 PO. Histopathological observation further confirms the radiological findings. No physiological side effects and other complications are observed in any group.

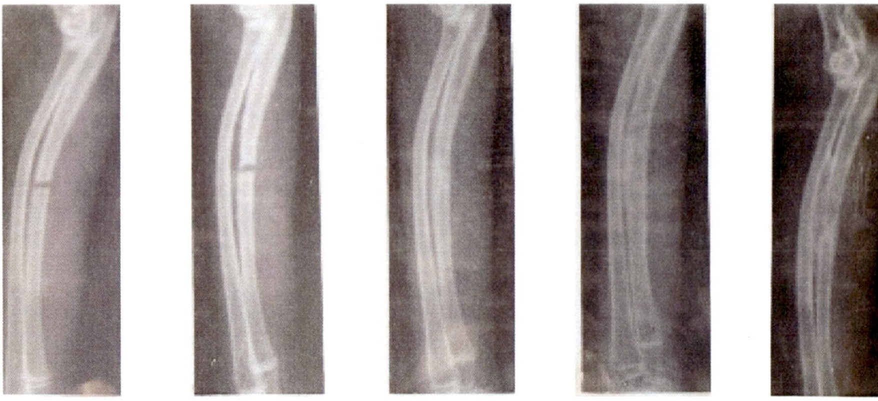

Day 0 10 20 30 40

Group A

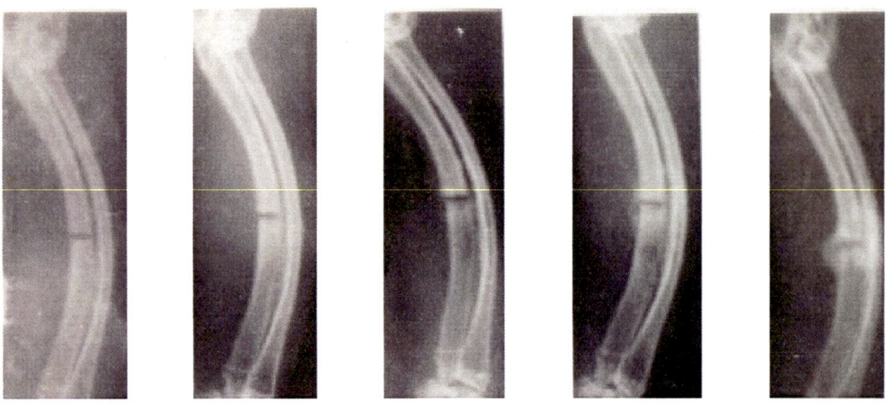

Day 0 10 20 30 40

Group B

Lateral radiograph of test group A showing accelerated fracture healing, group B (control) showing relatively delayed fracture healing in different time intervals

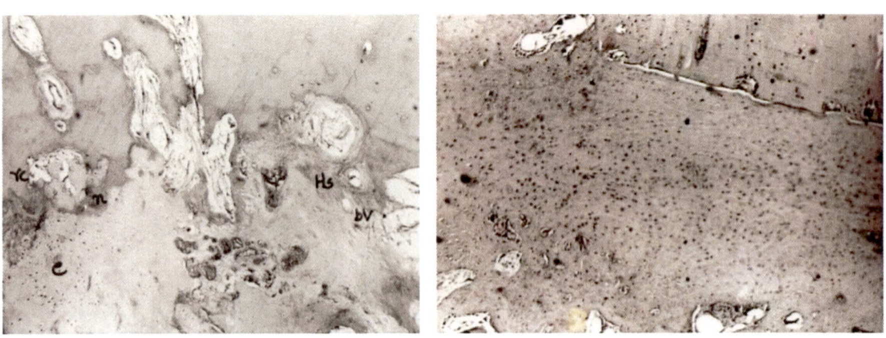

Histopathological photomicrograph of test group (left) showing abundant new bony trabeculae crossing over the fracture site, group B showing well visible fracture line at day 40PO

17

Evaluation of Herb, Cissus Quadrangularis (Hardjore) on Fracture Healing

India is bestowed with an enormous wealth of medicinal plants and has a rich heritage of treating human and animal ailments by using herbal medicines. However, there is need for a systemic, scientific approach to validate the authenticity of claims on plant-based medicaments. This approach has paid rich dividends in developing cheaper, safer, effective and eco-friendly remedies, which recently has gained worldwide attention. *Cissus quadrangularis,* Linn (Syn. *Vitis quadrangularis,* Wall; *Heliotropium indicum,* Linn) is a tender climber with fleshy quadrangular stems and grows in the hotter parts of Indian subcontinent, Sri Lanka and East Africa. In Hindi it is known as hardjore; in English as adamant creeper, bonesetter; in Ceylon as hiressa; in China as di xanh voung; in Burma as shazam-lese and in Bangladesh as harbhanga. Considering the advantage of herbal medicines, it is planned in this chapter to investigate the efficacy of *Cissus quadrangularis,* Linn, in one of the most commonly encountered problem of fractures in small animals.

Dried ethanolic extract of *Cissus quadrangularis* (CQ) is obtained from 500g of dried finely powered stems using soxhlet apparatus. It is dissolved in a vanishing base to make a semi-solid paste for local application. For parental use, this extract is reconstituted with distilled water (100 mg/ml) and filtered.

Twelve clinically healthy mongrel adult dogs aged 2-3 years and weighing 10-15 kg is randomly divided into two equal groups (A and B). In all animals operative site is prepared for aseptic surgery. Premedication is done with atropine sulphate @ 0.06 mg/ kg body weight subcutaneously and anaesthesia is achieved with a combination of Xylazine hydrochloride @ 1 mg/kg body wt. and Ketamine hydrochloride @ 10 mg/kg body wt. intravenously. Unilateral comminuted diaphyseal femoral osteotomy is created under general anaesthesia and the bone defect is immobilized with neutralization bone plating (AO/ASIF technique). All the animals are administered 5 mg/kg b wt Enrofloxacin intramuscularly for 5 d and 1 mg/ kg b wt Diclofenac sodium injection intramuscularly for 5 d postoperatively. Skin sutures are removed after 10 days. In the animals of group

B, in addition to the postoperative treatment, the ethanolic extract of the herb, *Cissus quadrangularis* is applied on the skin twice daily along with subcutaneous injection (50 mg/ kg body wt.) on every alternate day for 20 d postoperatively.

The operated limb of all animals is observed for the following parameters on days 1, 3, 7, 15, 30, 45 and 60 after surgery: (a) wound condition (b) degree of swelling (c) postoperative pain on gentle pressing at fracture site (d) weight bearing of the operated limb. On walking it is graded as test limb not touching the ground, toe of test limb touching the ground occasionally, touching the ground frequently, paw touching the ground with partial weight bearing and touching the ground with full weight bearing, (e) appetite, defecation, urination and general health of all the animals are observed throughout the period. Respiration rate, pulse rate and rectal temperature of all animals are also recorded before the start of the experiment and on days 1, 3, 7, 15 and 30 postoperatively.

Venous blood samples (7 ml) are collected in all animals before and on day 1, 3, 7, 15, 30, 45 and 60 postoperatively. Out of this, 5 ml is used for serum collection and remaining 2 ml is collected in a vial containing anticoagulant EDTA. Freshly collected blood is analyzed for total leukocyte count and differential leukocyte count. The serum samples are analyzed for the estimation of calcium, phosphorus and serum alkaline phosphatase. Oblique radiographs of the fracture site are taken just after surgery and subsequently on days 15, 30, 45 and 60 post-surgery. Radiographic factors are KVP=65, mAs=20 and FFD=90 cm.

The animals are sacrificed on day 60 and the osteotomy site is observed for callus size, status and position of metallic implants, fixation failure, if any and condition of surrounding soft tissues. Bone specimen from healing site and adjacent normal bone are collected and processed for (a) scanning electron microscopic study and (b) decalcification in Goodling and Stewart's solution followed by H&E staining.

Animals of both groups are maintained their normal appetite, defecation and urination. The surgical wounds is healed by first intention within 8-10 d in all animals except one animal each of group A and B, where suture dehiscence have occurred and the wound is healed by second intention in 12-15 d. Muscle atrophy is not apparent in any of the animals. Animals of group B shows very mild inflammation from day 3 upto day 15 postoperatively with moderate inflammation on day 7. In group A, inflammation is noticed up to day 20 postoperatively. Edema and pain are followed a similar course as that of inflammation.

Weight bearing is satisfactory in all the animals on the first postoperative day indicating good reduction and fixation. However, after first day, the weight bearing is gradually reduced due to pain and swelling. It improves rapidly in the animals of group B compare to A, due to early resolution of all inflammatory signs in this group of animals probably due to the presence of analgesic and anti-inflammatory properties of C Q, which has a steroid active principle. An early increase in the rectal temperature, pulse and respiration rates in animals of both groups are observed, which subsequently reduces towards normal values within 10 d postoperatively. Leukocytosis with neutrophilia and concomitant decrease in lymphocyte count is observed during early postoperative period in animals of both groups. The values subsequently are returned towards normal after 10 d. Leukocytosis, due to neutrophilia has been reported in conditions where there is corticosteroids release as in states of stress, pain, trauma and surgical manipulation.

There is an increase in the serum alkaline phosphatase (AP) activity immediately after surgery in both groups. However, the enzyme activity is further increased from day 10 onwards in the group B, up to the end of observation period. This increased AP activity may be suggestive of an increase in the osteoblastic proliferation during osteotomies repair. Serum calcium values are non-significantly increased immediately after surgery, followed by significant decrease by day 30 in group A, whereas significantly low levels of serum calcium on day 7-15 is noticed in group B. Decrease in the level of serum calcium to a greater extent in the CQ treated group (B) may be due to faster healing process with more mobilization of calcium in the formation of bridging callus. Low levels of serum calcium in CQ treated animals during bone healing is also reported by different workers and postulated that parafollicular or "C cells of thyroid gland release more calcitonine and thereby a decrease in serum calcium level since calcitonine increases the osteoblastic activity. Serum phosphorus level shows an increasing trend immediately after bone fixation in both groups and it remain slightly above normal level till the end of observation period.

Radiographs taken immediately after surgery, reveals good fixation reduction and alignment of osteotomized bone fragments in animals of both groups. Fixation is maintained and good bone healing is observed in all animals of group A and B except in one animal of group A in which bending and loosening of screws is observed. However, better bone healing with early remodeling is observed in the animals of group B. Bone plating with interfragmentary compression screws (neutralization plating) provides the most stable reduction and is especially indicates in comminuted femoral diaphyseal osteotomies reduction. This neutralization plate provides rigid fixation in these animals and shows better healing pattern. In one animal of group A bending and loosening of screws is observed, which may be due to the fact that there is a gap between the bone fragments and between the plate and bone which is susceptible to failure due to bending and torque.

Soft tissue healing is completed and the stainless steel plates are encapsulated by a fibrous covering in group A while in group B plates are deeply embedded within the newly formed bony callus. In histopathological examination in the group A, incomplete bone union is seen characterized by newly formed bony matrix with resorption cavities, budding blood vessels and Haversian system which is noticed in an uneven direction. Abundant bone marrow is present with few osteoclasts near the bony spicules. Periosteum is detached and thickened. Occasionally, soft callus is seen at the endosteal side with a marked fibrocellular reaction. Blood vessels and mononuclear infiltration is also evident within the marrow tissue. In the herbal group (B) complete union at the osteotomies site is evident. Osteoblastic and osteoclastic activity is marked at many areas. In these sections well-organized osteoid seam, Haversian systems and calcified cartilaginous tissue are evident. Periosteum shows a pronounced fibrocellular reaction within the newly formed bony tissue. There is also evidence of fully developed bony trabeculae, which cross over the osteotomies line. The marrow tissue is abundant and shows presence of osteoclastic activity. Blood vessels and mononuclear infiltration are also evident within the marrow tissue.

In scanning electron microscopic examination, in the group A, the defect is partially filled by newly formed osseous tissue. At certain areas, the collagenous network is relatively more and newly formed bone appears less compact, whereas in other areas, the osseous cells are clustered and indications of vascularisation and lying down of bone matrix. Further magnification shows the presence of vessels and bone matrix, which is in the process of calcification. In the herbal group (B), complete union and excellent remodeling at the osteotomies site are observed. The osteoclastic activity is marked and trabeculae are consolidated and thickened with well-developed vascularity suggestive of acceleration of bone healing. There is abundance of marrow cells, osteons with presence of osteoclastic activity. The mineralization is relatively more compared to the other group.

In conclusion, all the sequences of bone healing events are hastened in animals, given local as well parenteral ethanolic extract of Hardjore (Cissus quadrangularis) as an adjunct to internal fixation of comminuted diaphyseal femoral fracture in dogs. There are no any physiological complications observed in any animals of both the groups. No side effects of CQ are recorded in any animals of test group. It is concluded that *Cissus quadrangularis* accelerate the fracture healing.

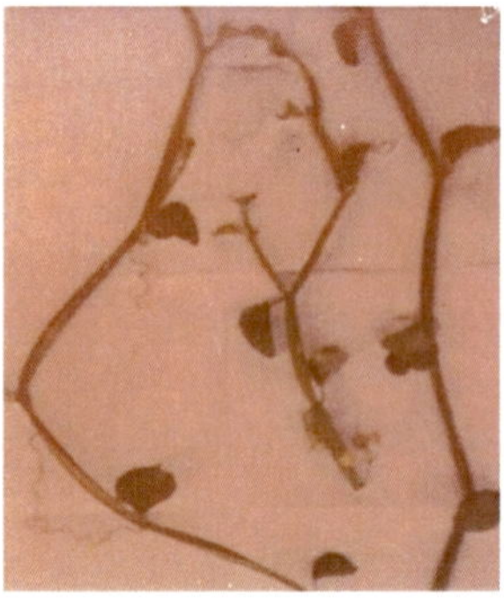

Herbal Hardjore (*Cissus Quadrangularis*) plant

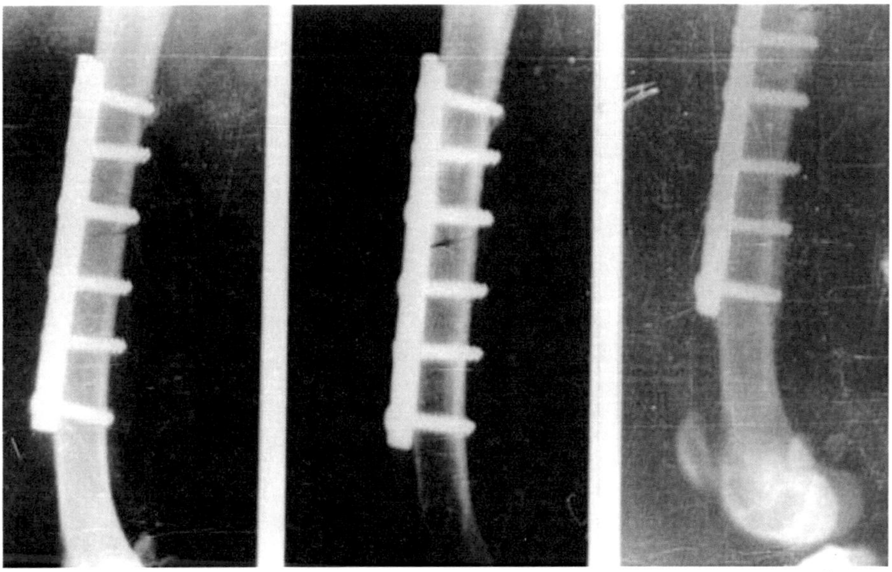

Group- A

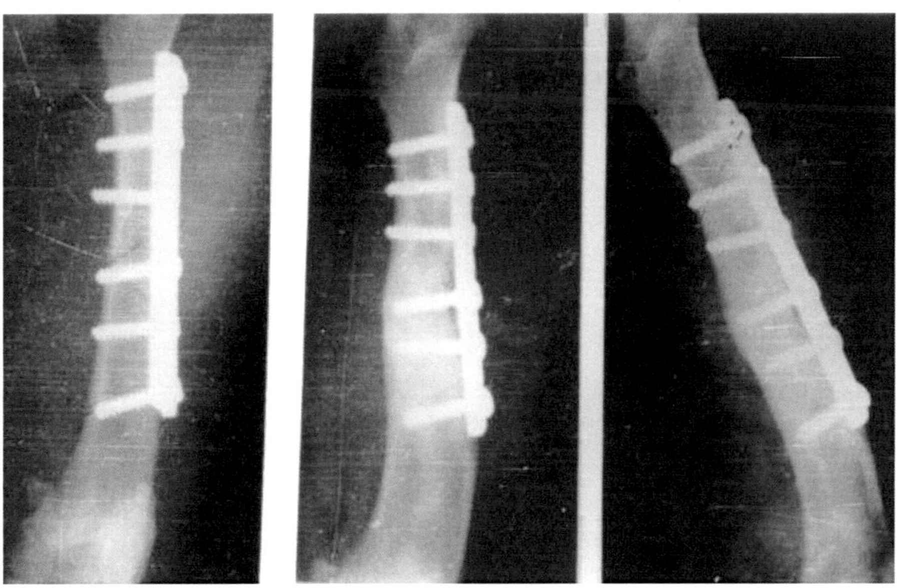

Group -B

Lateral radiograph of control group A (above) showing incomplete obliteration of osteotomies line and relatively delayed fracture healing, herbal test group B (CQ) showing accelerated and more intense fracture healing in different time intervals

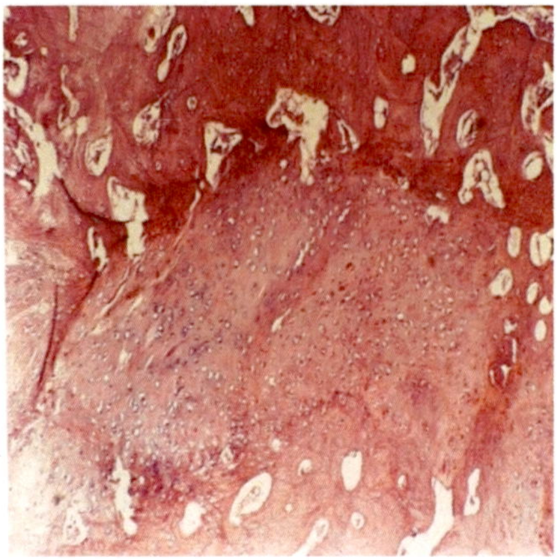

H& E photomicrograph (Gr A) showing little osteoblastic activity around the line of necrosis of cartilaginous matrix on both side of fracture

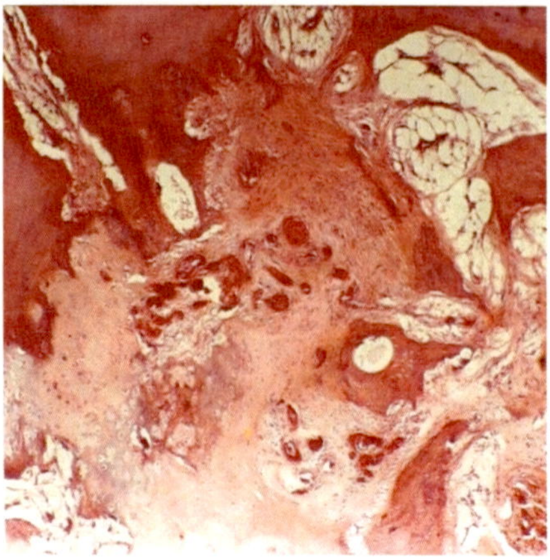

H&E photomicrograph of herbal group (B) showing well-organized osteoid seam, Haversian system and cartilaginous tissue at the site of fracture

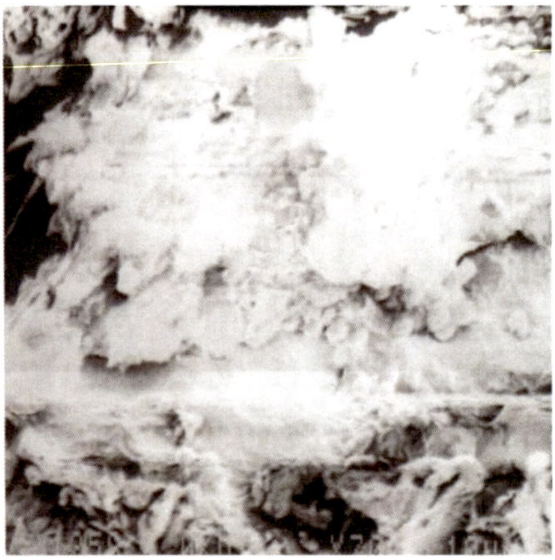

SEM photograph of Group A depicting cluster of bone cells within the mineralization matrix of the osteotomies site

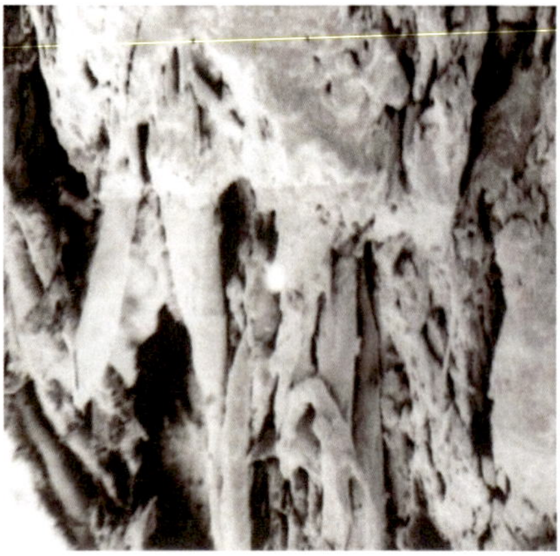

SEM photograph of herbal group (B) showing accelerated healing process along with the penetrating blood vessels at the osteotomies site

18

Role of Anabolic Steriod on Fracture Healing

Bone being a dynamic tissue undergoing constant biological changes, organic and mineral phases of fracture healing need an optimum general body metabolism and even a slight deviation from normal condition of the body is likely to cause considerable alterations in the healing process. Facture healing can be modified by variety of endogenous or exogenous factors that affect the metabolic function of cells. These factors include transforming growth factor beta, bone morphogenetic proteins, hormones, vitamins and prostaglandins. Anabolic steroids have the property of preventing the destruction and elimination of nitrogenous substances and minerals from the body. They help, thus, to have a positive nitrogen and mineral balance which is essential for healing and regeneration of any kind of tissue in a living organism. This chapter is undertaken to observe the effect of anabolic steroid (Durabolin) on fracture healing.

This study is conducted on ten clinically healthy adult New Zealand White rabbits of either sex divided into two equal groups (X and Z) consisting five animals each. All the animals are kept in standard diet and management throughout the experiment. In all animals, midshaft segmental defect (1cm) are created in both the ulnae under thiopental (2.5%) anaesthesia. In test group (Gr X), 10mg of Nandrolone phenyl propionate (durabolin) is injected intramuscularly in 5 days interval at 7 occasions. Lateral and anterio-posterior radiographs are taken on days 0, 7, 14, 21, 28 and 35 postoperative days. The radiographs are observed for progress of fracture healing and complications, if any, at different intervals.

Inflammation at the fracture site is increased upto 4th postoperative days in all the animals of both groups, and then the limb is normal in appearance by day 7. Surgical wound healing is by first intention in all the animals. Earliest return of weight bearing is seen in animals of test group followed by control. Body weight and food intake are increased two-fold in the test group of animals.

The lameness is seen in the immediate post operative period is attributed to surgical trauma and inflammation. Lameness is reduced gradually, which suggest that inflammation is subside and fracture is getting stable. The gain in body weight in the treated animals may be due to the anabolic effect of the steroid

(Durabolin) which produces positive balance of nitrogen and minerals in the body.

In test group of animals (group X) radiographs taken immediately after surgery shows radiolucent diaphyseal segmental defect in the ulna whereas radius acted as a natural splint. Both the fractured ends are rough. Radiographs taken at subsequent intervals (day 7 and day 14) show, that the bone defects become uniform. There are indications of new bone formation from the cut end of defect and the defect size is reduced and appears elliptical in shape. At day 21, further decrease in size of bone defect is seen. New bone growth is from both the cut end and obliterates the defect. Periosteal proliferation is also seen from the adjacent radius of bone defect. Radiographs on day 28, shows that the defect are almost completely filled with newly formed osseous tissues i.e. periosteal, endosteal and intercortical callus in seven limbs, whereas, in other three limbs, small area infiltrated with dense osseous tissue is seen. At day 35, bone defects are completely filled with newly formed woven bone in all 10 limbs. The periosteal new bone is well organized and in process of remodeling. Periosteal continuity is well developed at the fracture site. The medullary cavity across the bone defect is almost complete.

In control group (group Z) subsequent radiographs at day 7, 14 and 21 show fractured ends though become smooth and uniform but periosteal as well as endosteal proliferation along the defect is relatively slow as compare the test group (X). Radiographs of day 28 reveals new bone formation from both the cut ends as well as adjacent periosteal surface of radius but fracture gap is well visible. At day 35, the radiographs demonstrate that the activity of new bone formation is almost ceased and newly formed bone is in process of remodeling. Though the defect size is reduced but it is well visible at day 35 suggest delayed union. In this stage, periosteal proliferation, particularly from the proximal end, bridged with adjacent radius and periosteal continuity is also directed in this manner.

Repair of bone fracture involves the availability of building blocks in both organic and mineral phases of fracture healing. Sufficient organic materials in the early period and mineral salts in the later weeks largely depend upon the positive nitrogen and mineral balance after fracture trauma. If the natural sequences of negative balances of the above substances following fracture can be averted and a positive balance induced, one can expect a more intense and rapid healing. The anabolic hormone (durabolin) is used in this study in test group (X) has been able to intensify both the phases of healing. The acceleration of healing process in group X when compare with control (Z) is probably due to the anabolic effect of the steroid which produces a positive balance of nitrogen and minerals throughout the period of repair. Further, the gain in weight of the test animals receiving this hormone is so marked that anabolic effect of this hormone can be establish beyond doubt. Thus, it is obvious that anabolic hormones in fractured animals accelerate the total activity of healing process in all stages. This anabolic steroid treatment suggests that it promotes osteogenic activity and early calcification by allowing early calcium and phosphorus at fracture site. It is also reported that anabolic hormones causes exaggerated mucoplysaccharide response with an earlier mineralization of the fracture callus. It is also reported that anabolic steroid promotes skeletal growth and protein synthesis, and induces a positive nitrogen balance in the body, thereby retaining the essential minerals in the body and then accelerated the process of fracture healing which is also observed in this study.

In conclusion, Nandrolone phenyl propionate (Durabolin) when used as an adjunct, it accelerates different phases of fracture healing.

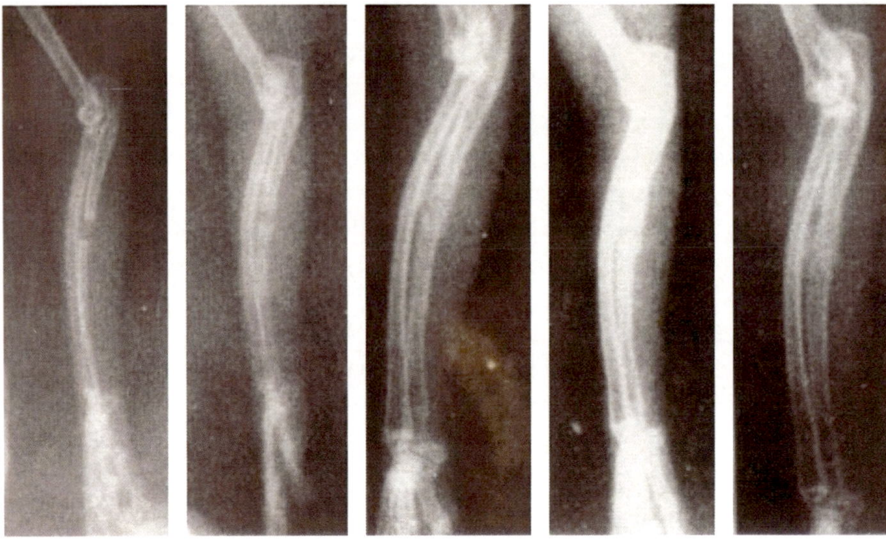

Day 0 14 21 28 35

Group X

Lateral radiographs of animals of test group showing accelerated course of fracture healing

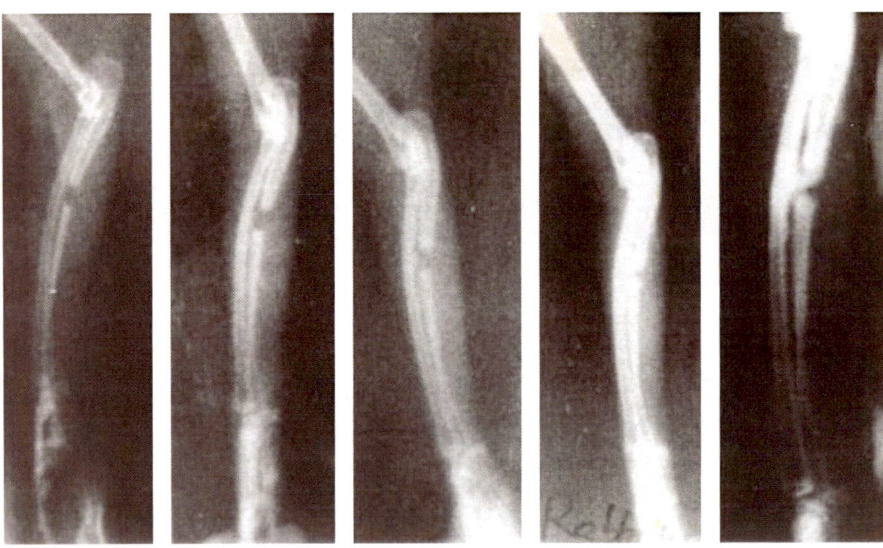

Day 0 14 21 28 35

Group Z

Lateral radiographs of animals of control group showing normal course of fracture healing

19

Role of Sex Hormone Progesterone on Fracture Healing

The action of the sex hormone is not confined to the growth and development of the sex organ alone, but the normal body growth as well as the growth of the skeleton are to a great extent influenced by these hormones. It is well documented that following oopherectomy, female experience a "rapid phase" of bone loss for the next few years. Of the female gonadial hormones, the action of estrogen on skeletal growth has been studies by various workers and shows that estrogen replacement therapy can help to prevent the postmenopausal stage of accelerated rate of bone loss. The effects of progesterone on bone healing or bone growth has not yet been studies in detail. The objective in this chapter is thus undertaken to investigate the effects of progesterone supplement on cortical bone healing in spayed rabbits.

The study is conducted on ten clinically healthy female adult New Zealand White rabbits divided into two equal groups (A and B) consisting of five animals each. All animals are spayed and kept in standard diet and management throughout the experiment. In all animals, midshaft simple transverse fractures are created in both the ulnae under thiopental (2.5%) anaesthesia. In test group (gr A), 1mg (=1ml) of progesterone is injected subcutaneously in alternate days for upto 35 post fracture days. Lateral and anterio-posterior radiographs are taken on days 0, 7, 14, 21, 28 and 35 post-fracture. These radiographs are observed for progress of fracture healing and complications, if any, at different intervals.

The specimens for electron bone scan are collected from the healed site of test bone immediately after euthanizing the animals on day 35. Every specimen is consisted of both healing site and adjacent host bone. They are fixed with 5% glutaraldehyde phosphate buffer solution for 24-48 hr. After fixing in glutaraldehyde, two washes for 30 min each specimen with phosphate buffer saline (pH 7.4) followed by distill water is performed. Then the specimen is dehydrated in serial concentration of ethanol 30-100% for 30 min. Thereafter, the specimens are dried with Hexamethyldisilazine (HMDS) for 20 min. After the golden ion sputtering is done on Jeol ion sputter Model (JFC 1100) at 7-10 mA and 1-2 KV for 5 min. Finally, they are examined in scanning electron microscope (Jeol ISM 840 model) for the study of their different surfaces to identify the orientation and distribution of newly formed osseous tissue at fracture site.

In test group (Gr A) of animals, the series of radiographs taken at different intervals, shows mid-diaphyseal transverse fracture of ulna. The reduction and alignment is maintained throughout the period of observation. On day 0, the fracture fragments are either perfectly aligned or have minor deviation. This is, however, maintaining in subsequent intervals. On day 14, there is extensive periosteal reaction extending to almost entire diaphysis of the ulna and also in the adjacent radius. This periosteal reaction appears more pronounce on day 21. At this stage, the fracture gap is almost filled by newly formed osseous tissue and bridging of periosteal callus is complete. On day 28, there is marked reduction in periosteal reaction, bridging callus is organized and osseous union is complete. The osseous union between radius and ulna adjacent to fracture site is also appreciable. On day 35, remodelling of healing site can be appreciated by marked reduction in callus size and smooth, uniformly dense periosteal surface. Periosteal continuity is well developed at the fracture site, however, not complete.

In control group (gr B), the reduction and alignment of fracture fragments are also maintained throughout the period of observation. On day 0, the fracture line is complete and transverse to the longitudinal axis of ulna. On subsequent intervals (day 7 and 14) there is marked increase in the gap between the two fragments and there is no indication of periosteal reaction. On day 21, the fracture gap appears irregular. There are thickening of periosteal surface adjacent to fracture line suggesting of periosteal reaction. This becomes more pronounced on day 28 post-fracture and there is indication of a small organized callus formation, however, the fracture gap is still visible. This periosteal as well as intercortical callus becomes more organized and it's bridged across the fracture line on day 35 postoperatively. The facture gap though becomes hazy yet it can be distinctly visualized. In general, fracture healing is relatively delayed in this group as compare to test group.

The scanning electron microscopic (SEM) observations provide the status of fracture healing in both groups. It also provides topographic picture about the type of tissues involves in the healing and status of vascularity at the fracture site. In animals of test group (Gr-A), SEM clearly indicates osseous union at the fracture end, however, process of remodelling is still continuing. This is evidenced by the presence of bone cells (osteoblast) and un-mineralized collagen network at the defect site. Further magnification of the section shows that the healing site appears hyper-vascularized and new vessels is mostly from the periosteum and surrounding soft tissues. In animals of control group (Gr-B), the SEM picture shows that the bone defect is partially filled with fibrocartilagenous tissues. The osseous tissue is predominantly seen along the fracture ends, whereas the fibro cartilage is present at the center of the defect. The process of lying down of bone matrix and vascularisation is also relatively less in this group of animals.

It's postulated that progesterone supplementation will prevent the loss of bone mass following the menopause in females in a manner similar to that of estrogen. It is also suggest that progesterone supplementation of the spayed dam resulted in an enhancement of individual bone cell activity and a shortening of the functional life-span of the osteoclast population. Progesterone supplementation alters cortical bone remodelling activity in the absence of ovarian hormones. This alteration is manifested by an increase in activation frequency, the resulting basic multicellular units (BMUs) having a markedly shortened resorption time and a shortened bone formation time. This may explain to the prevention on bone loss in the early post-menopausal period of the female when she is given supplementary progesterone. Remodelling time in test group is also reduced may be due to the early activation of osteoclasts. The exact mechanism of action of this hormone in bone remodelling is still unknown but one would have to consider the possibility of competitive bindings of progesterone for glucocorticoids receptors on bone cells.

Conclusion

Progesterone supplementation enhances fracture healing process in spayed animals or in the absence of ovarian hormones.

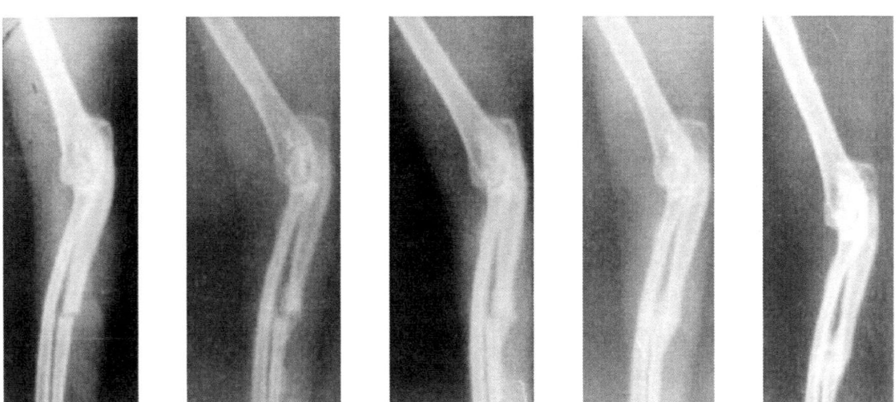

Day 0 14 21 28 35

Group A

Lateral radiographs of animals of test group showing accelerated course of fracture healing

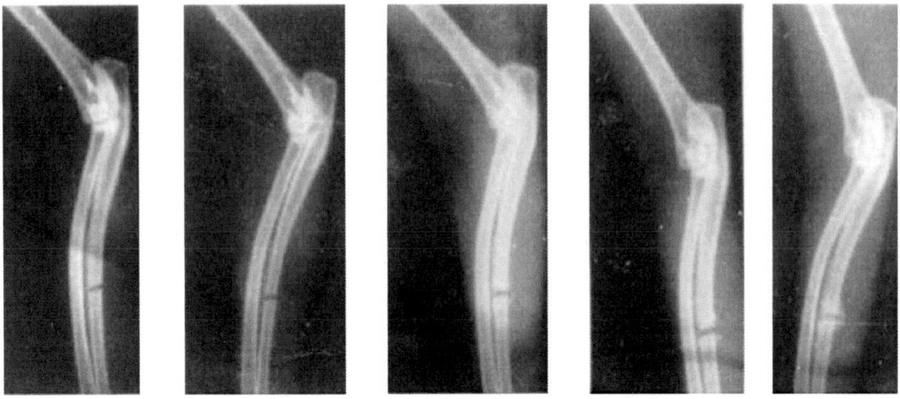

Day 0 14 21 28 35

Group B

Lateral radiographs of animals of control group showing normal course of fracture healing

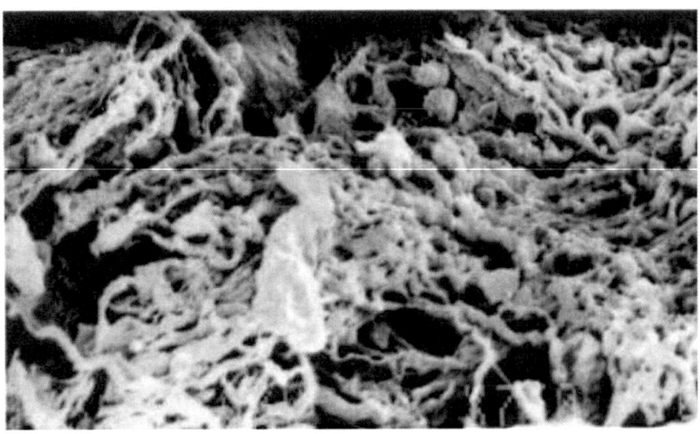

SEM photograph of group A showing laying down of bone matrix-collagen network and fusion with newly formed bone cells-osteoblast

SEM photograph of group B showing new bone formation at the fracture ends only and center is filled with fibrocartilagenous tissue

20

Effect of Ultrasound Therapy on Fracture Healing

Bone healing can be modified by variety of endogenous or exogenous factors that affect the metabolic function of bone cells. The Food and Drug Administration is approved the use of low-intensity ultrasound for the accelerated healing of fresh fractures in October 1994 and for the treatment of established nonunion in February 2002. The first regulatory approval is based primarily upon two rigorous, double - blind, placebo controlled clinical trials, which shows that the rate of healing of fresh fracture is accelerated by treatment with ultrasound. Therapeutic ultrasound has been able to improve functional activity, reduce post-ambulatory period, alleviate suffering and restore the animal's health early. It has been used frequently for acceleration of healing in various musculotendinous injuries in animal. However, the role of ultrasound therapy in different dose levels in fracture healing has not been elucidated properly. In view of the above fact, this chapter is undertaken to investigate the effects of therapeutic ultrasound in two different doses, on cortical bone healing in rabbits.

Twelve clinically healthy adult New Zealand white rabbits of either sex are divided equally into three groups A, B and C, and kept on standard diet and uniform management. In all animals, mid-shaft simple transverse fracture is created in both the ulnae under thiopental (2.5%) anaesthesia. All the animals are administered broad-spectrum antibiotic and analgesic and surgical wounds are dressed daily with povidone iodine and antibiotic cream for 4 days postoperatively. Group A animal is received pulsed ultrasound therapy @ 1W/cm^2 (1:1) whereas, group C animals is received continuous ultrasound therapy @ 3 W/cm^2, starting on day 4 after the creation of fracture. Ten treatments, each consisting of 5 min duration, are given on alternate days. Group B animals are served as control. In both the test groups (A and C), the transducer head of the ultrasound machine is moved in a linear fashion along the length of the bone after applying ultrasound gel.

All the animals are evaluated clinically for the extent and duration of trauma on the basis of various parameters i.e., general examination in pre and post-ultrasound therapy, wound condition, degree of swelling, postoperative pain, weight bearing of the operated limb and lameness (if any), rectal temperature, feeding pattern and general behavior on days 0, 1, 3, 5, 7, 9, 15, 30 and 40 post-fracture. Improvement of any condition is evaluated by the following score:

Nil : No improvement;

\+ : Detectable improvement but condition not resolved;

++ : Condition fully resolved by the end of treatment;

+++ : Rapid and total restoration.

Approximately 4 ml of arterial blood is collected from the middle ear artery, after placing the rabbit in a restrain box on days 0, 1, 3, 7, 9, 15, 30 and 40 post-fracture and analyzed for haemoglobin (Hb), total leukocyte count (TLC) and differential leukocyte count (DLC). Lateral and anterio-posterior views of the fracture site are taken just after surgery and subsequently on days 10, 20, 30 and 40 post-surgery to observe progress of fracture healing and complication (if any). The radiographic data are KVP= 65, MAS=20 and FFD=90 cm. The specimens are collected from the healed site of test bone immediately after euthanizing the animals on day 40. Every specimen is consisted of fracture healing site with adjacent host bone. The bone sections are preserved in 10% formalin for 9-10 days and then are decalcified in Goodling and Stewart's solution (15% formic acid, 5% formalin and 80% distilled water). Completely decalcified bone sections are then processed routinely for H&E staining. These sections are observed for status of healing at fracture site.

Animals of all the groups are started taking feed and water partially within 24 h of operation, which became normal by 4th postoperative day. Clinical observations reveal significant increase in rectal temperature up to day 3 postoperative may be due to surgical stress and inflammation. Surgical wounds in all the animals are healed by first intention with negligible scar.

Inflammatory swelling at the fracture site is increased gradually and it is moderate on day 4 in all the animals. In group B, it is moderate to severe on day 6 and moderate to mild from day 8-15 post-fracture. In group A, the swelling is moderate to mild from day 6 to 9. In group C, the swelling is moderate to severe from day 6 to 20 and severe to moderate from day 22 to 30. Pain is moderate to severe on day 4 in all the animals. It is remained severe on day 6 and moderate from day 8 to 10 in the control animals. In group A, pain is moderate on day 6 and mild on day 10 and after that it is absent whereas, in group C, pain is severe from day 6 to 16 and then moderate to mild from day 18-20. Weight bearing is followed the same pattern as inflammatory swelling and pain. Earlier return of weight bearing is seen in animals of group A followed by groups B and C. Among the two treatment groups, improvement of function is far better and earlier in group A than group C.

Values of Hb, TLC and DLC does not vary significantly at different intervals in all the three groups of animals except significant neutrophilia is noticed upto day 7 in groups A and B and upto day 15 in group C animals. The neutrophils have been recognized as the first cellular defense of the body and neutrophilia along with leukocytosis is generally associated with inflammation. In group C, inflammation is severe for long duration, which may have causes significant neutrophilia even day 15 post-fracture.

In group A animals, radiographs taken immediately after surgery shows radiolucent diaphyseal transverse defect in the ulna, whereas, radius acted as a natural splint. On day 0, the fracture fragments are rough and perfectly aligned. On day 10, periosteal reaction is started from both the fracture ends, which become uniform in appearance. Though, the fracture gap is slightly increased as compare to day 0, but gap started obliterated by newly formed periosteal proliferation, which appears more pronounced on day 20. At this stage, the fracture gap is almost filled by newly formed osseous tissue and bridged the defect. However, a thin fracture line is still visible at this stage. On day 30, there is marked reduction in periosteal reaction as bridging callus is organized and osseous union is completed. The osseous union between radius and ulna adjacent to fracture site is also appreciable. Periosteal continuity is well advanced. On day 40, marked reduction in callus size and smooth, uniformly dense periosteal surface can appreciate remodeling of healing site. Periosteal continuity is well developed at the fracture site. The union of medullary cavity across the facture line, however, is not complete.

In group B, subsequent radiographs of days 0, 10 and 20 shows that fracture ends though become rough and uniform but periosteal as well as endosteal proliferation along the defect is relatively slow compared to the treated group A. Radiographs of day 30, reveal new bone formation from both the cut ends as well as adjacent periosteal surface of radius, but fracture gap is well visible. At day 40, these periosteal proliferations are more intense. Though the defect size is markedly reduced but it is well visible at day 40 post fracture suggesting delayed healing compared to group A.

In animals of group C, radiograph taken on day 10, both the radius and ulna become fractured at the original site of ulnar defect and both the bones are slightly deviated from their original position. However, all the fractured ends are aligned with their corresponding ends. Soft tissue swelling is marked at this stage and there is no sign of any periosteal reaction in any of the bones. Radiographs of day 20 reveal that both the fractured bones distracted with each other and proximal end of radius is aligned with distal end of ulna, whereas proximal end of ulna is deviated laterally from its corresponding end of ulna and there are no signs of any periosteal reaction. However, periosteal reaction is clearly visible in the radiograph of day 30. This reaction is started from all the four fractured ends of radius and ulna and developed as a bridging callus among these ends at the fracture site. At this stage, soft tissue swelling is minimal. On day 40, this bridging callus is more intense and smooth and uniform. Periosteal continuity is well developed in both the fractured bones, which healed as "mal-union." However, the union of medullary cavity across the facture line is not complete and remodelling process is yet to be started.

The fracture healing appears to be radiographically better in the treated animals (group A) compare to control. It is reported that doses of high intensity ultrasound and over treatment have resulted pathological fracture and sub-periosteal damage which is also found in this study in the animals of group C, where higher dose ($3W/cm^2$) is used. Low intensity ultrasound has demonstrated an acceleration of bone healing and more profound callus formation in animal and human clinical experiments as similar to results in the animals of group A.

Histopathologically, in the animals of treated group A complete union and excellent remodelling at the fracture site is observed. In group B, incomplete union at the fracture site is evident and the fracture gap and newly formed bone are well distinguished from that of normal bone. The newly formed osseous tissue near normal bony tissue is evinced in an improper direction and is not fully developed and does not cross over the fracture line. Blood vessels and mononuclear infiltration is also evident within the marrow tissue. In group C, two animals shows delayed union with a relatively large fracture gap filled with fibro cartilaginous connective tissue. Periosteal healing is also evident. The cartilage is seen arising from the cambium layer of the periosteum and is replaced by bone near the fracture site. Excessive chondroblastic and osteoblastic proliferation with abundant marrow tissue is seen along with osteoclastic activity. In two animals, non-union is observed with fibrous tissue filling the fracture gap. In some sections, though there is evidence of bony trabeculae, which, however, are not fully developed and does not cross over the fracture line. These histopathological results confirm the clinical and radiographical findings.

At lower frequencies, therapeutic ultrasound can penetrate more deeply. In addition to its thermal effects, ultrasound causes micro-massage. The mechanical reaction causes biochemical changes in the tissues with improvement in blood supply, lymph flow and intracellular metabolism. In group C, where higher frequency of ultrasound ($3W/cm^2$) is given, a prolonged severe inflammation occurred as higher frequency penetrates less effective distance and causes superficial soft tissue damage and inflammation. The shear stress waves are created increase the local heat and since the periosteum is avascular, no cooling occurs and the excessive heat will cause pathological fracture and severe pain, which is also finding in this study. In addition to modulating gene expression, ultrasound may increase blood flow through the dilatation of capillaries and the enhancement of angiogenesis accelerating fracture healing. It is also reported that low-intensity ultrasound reduced the healing times of fresh fractures by up to 40% and its exposure in animal model of the endochondral process.

In **conclusion**, lower frequency of ultrasound accelerates fracture healing process, whereas, higher frequency of ultrasound damage the bone and retard healing process.

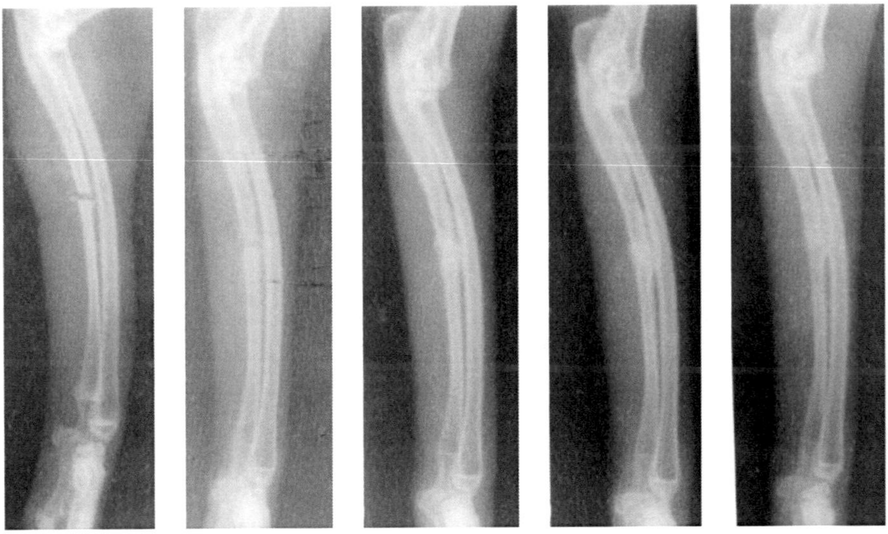

Day 0 10 20 30 40

Group A

Lateral radiographs of test group A showing accelerated course of fracture healing

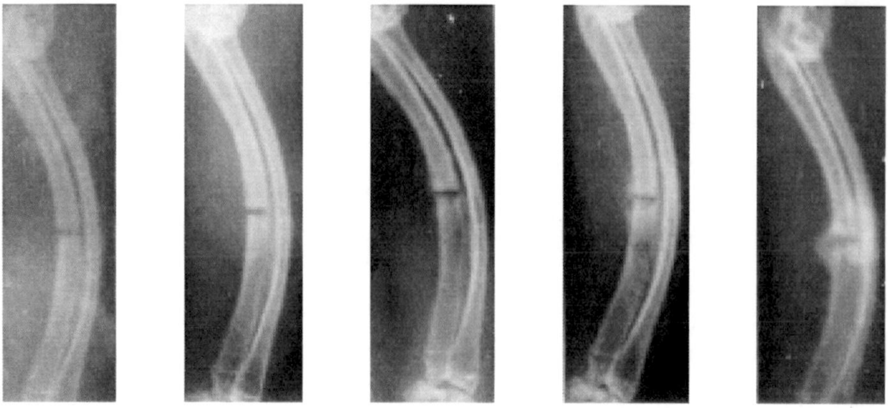

Day 0 10 20 30 40

Group B

Lateral radiographs of test group B showing normal course of fracture healing

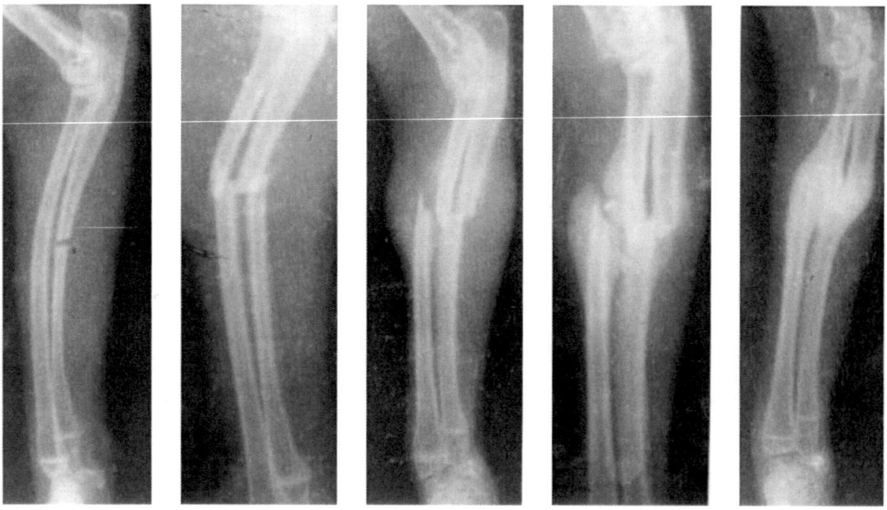

Day 0 10 20 30 40

Group C

Lateral radiographs of test group C showing mal-union type of fracture healing

21

Effect of Non-Steroid Anti-Inflammatory Drug (NSAID) on Fracture Healing

Standard postoperative management of animals with fractures includes the use of nonsteriodal anti-inflammatory drug (NSAID). Nonsteriodal anti-inflammatory drug inhibits inflammatory responses by blocking prostaglandin (PG) synthesis via the Cyclooxygenase pathway of arachidonic acid metabolism. However, PG is important factor in early phase of bone healing. A local increase of PG concentration is a response of bone to trauma and PG (particularly PG_2) may stimulate differentiation and proliferation of osteoprogenitor cells during early bone healing. It has been reported that NSAID, such as Diclofenac acid, ibuprofen, indomethacin, aspirin, salicylic acid decreased ingrowths of bone into porous coated implants and decreased mineral content in rabbits. In view of the above fact, this chapter is undertaken to investigate the effects of one potent NSAID (Flunixin Meglumine) on fracture healing.

The study is conducted in ten clinically healthy adult New Zealand White rabbits of either sex divided into two groups (A and B) consisting of five animals each. In all animals, mid-shaft simple transverse fracture is created in both the ulnae under thiopental (2.5%) anaesthesia. In animals of group A, 0.25ml (12.5 mg) Flunixin Meglumine (Ilium Flunixil) is injected intramuscularly on alternate days from day 5-20 postoperatively whereas, in group B no drug is given. Lateral and anterio-posterior radiographs are taken on day 0, 10, 20, 30 and 40 post-fracture. The radiographs are observed for progress of fracture healing and complication, if any, at different interval. The specimen for histopathological study is collected from the healed site of test bones immediately after euthanizing the animals on day 40. Every specimen is consisted of both healing and adjacent host bone. The bone sections are preserved in 10% formalin for 9-10 days and then are decalcified using Goodling and Stewart's solution (15% formic acid, 5% formalin and 80% distilled water). Completely decalcified sections are then processed routinely for H&E staining. These sections are observed for the status of healing at the fracture site, at cellular level.

Inflammation at the fracture site is increased upto 4th postoperative day in all animals of both groups, them the limb is normal in appearance by day 7. Surgical wound healing is by first intention in all animals. Earlier return of weight bearing is seen in animals of test group followed by control.

In test group of animals (group A) radiographs taken immediately after surgery shows radiolucent diaphyseal transverse defect in the ulnae whereas radius acted as a natural splint. On day 0, the fracture fragments are perfectly aligned. Subsequent radiographs on day 10, 20 and 30 shows fracture gap gradually increases, fracture ends become smooth and periosteal as well as endosteal new bone proliferation along the defect is negligible compare with the control group (group B) in the same interval. Day 30 radiograph shows some periosteal reaction started from the cut ends and adjacent radius. On day 40, both the cut ends appears dense and the bone defect appears relatively more cloudy and hazy in comparison to that on day30, due to some periosteal reaction trying to bridge the defect. However, fracture gap clearly visible and bony union not takes place at this stage. No signs of any infection are observed in any animal at any interval. In control group (B), periosteal reaction is started as early as on day 20. On day 30, this periosteal reaction is extending to almost entire diaphysis of the ulna and also in the adjacent radius reducing the fracture gap. On day 40, bridging callus become more intense and organized and it is almost obliterating the fracture gap. In this stage, periosteal callus is also bridged with the adjacent radius adjacent to the fracture site.

Histopathologically, in the NSAID treated group (A), incomplete union at the fracture site is evident and the fracture gaps and newly formed little osseous tissue well distinguished from that of normal bone. This newly formed osseous tissue, which is mainly developed from periosteal end of the fracture, are not fully developed and crossover the fracture line. At the center of the defect, however, there is predominance of cartilaginous and fibrous tissue. Blood vessels and mononuclear cell infiltration is also evident in some places within the marrow tissue.

In group B (control) the fracture gap is almost filled with newly developed bony trabeculae, which is evinced by its proper direction and is woven /immature in nature. In some animal, there is some evidence of reorganization of this immature bone and development of new marrow and represented the total reparative process. Some newly formed thickened bony trabeculae penetrating from both the fracture ends are also noticed. The periosteum is also thickened and infiltrated with some mononuclear cells.

From the radiological and histopathological observations, it is revealed that the administration of NSAID appears to decreasing new bone formation at the fracture site, which causes delayed union. Several studies is recorded a drug and dose related effect of ketorolac, salicylic acid and indomethacin on fracture healing in rats and rabbits. In rabbits, indomethacin causes decrease bone mineral content, mechanical bending strength and blood flow at 2 weeks after osteotomy. It is also reported that indomethacin decreased Haversian remodelling in rabbits. It is reported that phenylbutazone decreased mineral apposition rate (MAR) in cortical bone and appears to decrease healing rate of cortical defects in rabbits, is similar to this study. The effect of this NSAID on bone activity may be due to the fact that non-steroidal-anti-inflammatory drugs inhibit inflammatory responses, which is to some extent, essential for activation of local and systemic autocrine, paracrine and hormonal mechanism in the body which promote recruitment and differentiation of osteoprogenitor cells in the periosteum and endosteum.

In **conclusion**, administration of NSAID drug retard fractures healing in experimental animals.

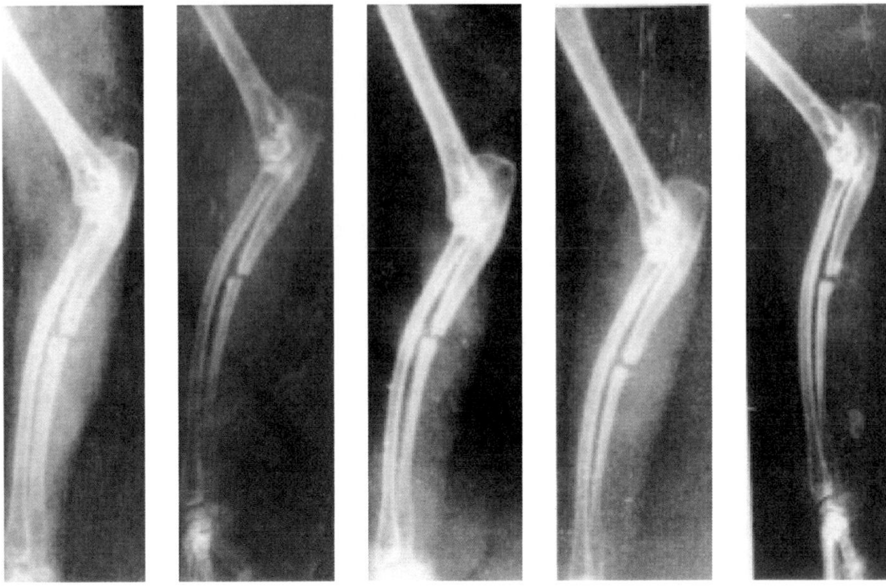

Day 0 10 20 30 40

Group A

Lateral radiographs of test group A showing retard fracture healing

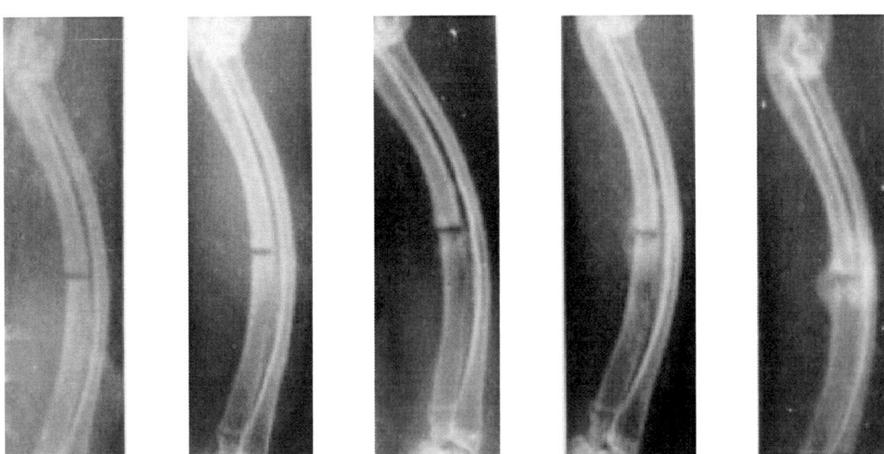

Day 0 10 20 30 40

Group B

Lateral radiographs of test group B showing normal fracture healing

22

Minimally Invasive Fracture Repair Techniques and Biological Osteosynthesis

Minimally invasive percutaneous osteosynthesis (MIPO), or minimally invasive fracture repair, was introduced into human orthopedic surgery during the 1990s because of several distinct advantages over traditional open reduction techniques. In MIPO, large incisions to facilitate open reduction are avoided and implants are introduced through small stab incisions in the skin. This led to advent to more biology friendly techniques. These biological techniques lay stress upon maximal preservation of blood supply around the fractured bone by minimal direct handling of the fracture environment. MIPO offer advantages which include preservation of the fracture hematoma; less surgical trauma to the surrounding soft tissues, reduced operative time therefore, decreased risk of infection and ultimately leading to indirect bone healing with abundant callus. Minimally invasive plate osteosynthesis (MIPO) is a novel technique for application of principles of biological fracture healing with a philosophy of dealing with soft tissues with utmost respect. Following advancements in implant technology and development of techniques of indirect fracture reduction this technique gained more popularity. The evolution of MIPO began with the use of bridge plating. Initially a conventional plate is applied for comminuted femoral fractures using long incisions but with preservation of the vastus lateralis muscle resulting in healing with abundant callus. Subsequently, the size of incisions got decreased, rather two small incisions are given at proximal and distal ends and plates are bridged sub muscularly over the fracture fragments. This technique causes minimal distress to soft tissues and bone, provides access to the bone through soft tissue windows, with minimal or no contact with the fracture by indirect reduction tools and leaving behind minimal foot prints.

Principles of biological osteosynthesis

The principles of biological osteosynthesis are developed in order to maximize healing potential by balancing biology and mechanics in the treatment of fractures. The basic principles of biological osteosynthesis include:

1. Minimize iatrogenic soft tissue disruption.
2. Utilize indirect fracture reduction techniques.
3. Provide appropriate stable fixation.
4. Promote the early return to limb function

The study focuses over various aspects of application of MIPO technique used in commonly indicated fractures, along with factors influencing its mechanical stability, recent advances and its downside. The advantages of MIPO include less soft tissue trauma, preservation of the periosteal blood supply and fracture hematoma, and decreased surgical site infection (SSI). Development of new implants including the less invasive stabilization system (LISS), point contact fixator (PC-Fix), and the locking compression plate (LCP) made indirect reduction and percutaneous stabilization much more practical.

Indirect reduction and MIPO of comminuted fractures is more feasible in human and small-animal veterinary orthopedics, where less rigid fixation techniques are often desirable. The large size of the horse requires maximal stability for fracture repair. Therefore, equine fractures amenable to MIPO are limited to inherently stable fractures. Nondisplaced, distal limb fractures in the horse are excellent candidates for MIPO because of the minimal soft tissue coverage and limited extraosseous vascularity. Smaller incisions likely reduce the risk of contamination during surgery, with decreased infection rates noted in MIPO of long bone fractures. LCPs are ideal for MIPO, as the stability of the construct does not rely on frictional forces between the plate and bone; therefore, plates can be applied without extensive soft tissue and periosteal dissection. This also means that plates do not require precise contouring, which facilitates minimally invasive placement. In conventional bone plating, stability is compromised when the load experienced by the construct exceeds frictional forces, causing some displacement of the plate along the surface of the bone. Additionally, the round head of the bone screws allow for minimal shifting within the oblong plate holes, a phenomenon that is overcome by the interlocking at a fixed angle between the LCP and locking head screws. It should be noted that although LCPs are better suited for MIPO, minimally invasive fracture repair is also achievable with standard dynamic compression plates (DCPs) and limited contact dynamic compression plates (LC-DCPs).

Indications of MIPO

Although minimally invasive approach for plating bears numerous advantages, due to its limitations it is not universally applicable. A careful preoperative planning is required prior to its implementation with bearing in mind the anatomical aspect of the fracture, condition of soft tissue, etiology of the fracture, type of construct required and date since injury.

- Epi-/metaphyseal fractures procedure
- Poor soft tissue condition obviating an open procedure
- Conditions where intramedullary nailing is not feasible
- Unsuitable fracture pattern (intra-articular extension), narrow, deformed or obstructed intramedullary canal
- Cases in which intramedullary reaming is contraindicated or can be detrimental
- Fractured bones with other implants or prosthesis *in situ*
- Settings where of an image intensifier is not available

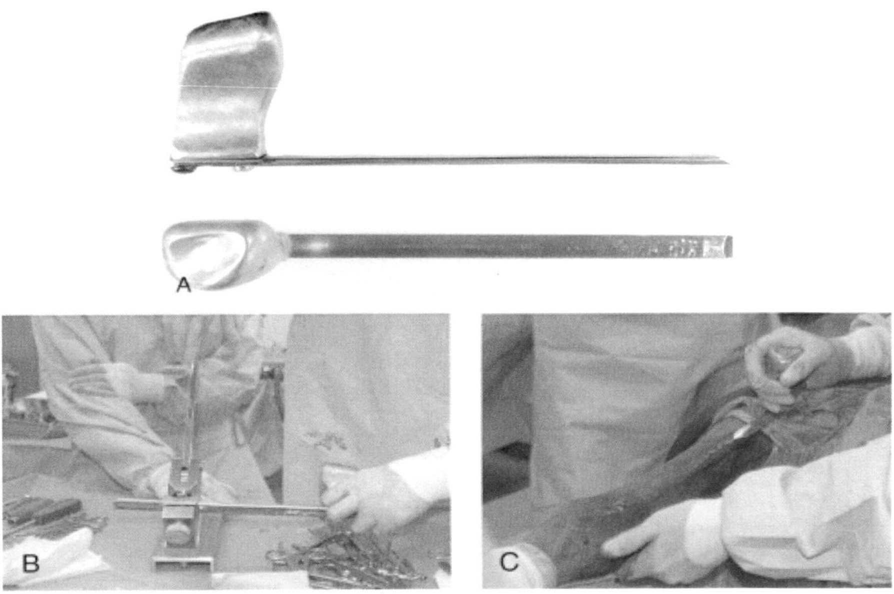

(A) A plate-passing device made from a broad DCP with a sharpened end and handle to facilitate formation of a subcutaneous tunnel. (B) The plate-passing device being contoured using a large plate-bending device. (C) The plate-passing device being introduced through a proximal incision. The device is used to create a subcutaneous tunnel on the dorsolateral surface of the third metatarsal bone.

An intraoperative image of a medial condylar fracture being repaired with a dorsolateral LCP placed minimally invasively. (A) The LCP is being slid into place using a long drill guide threaded into the distal hole. (B) The drill guides are threaded into the LCP through stab incisions. (C) The stab incisions are closed with simple interrupted sutures once all the screws are placed.

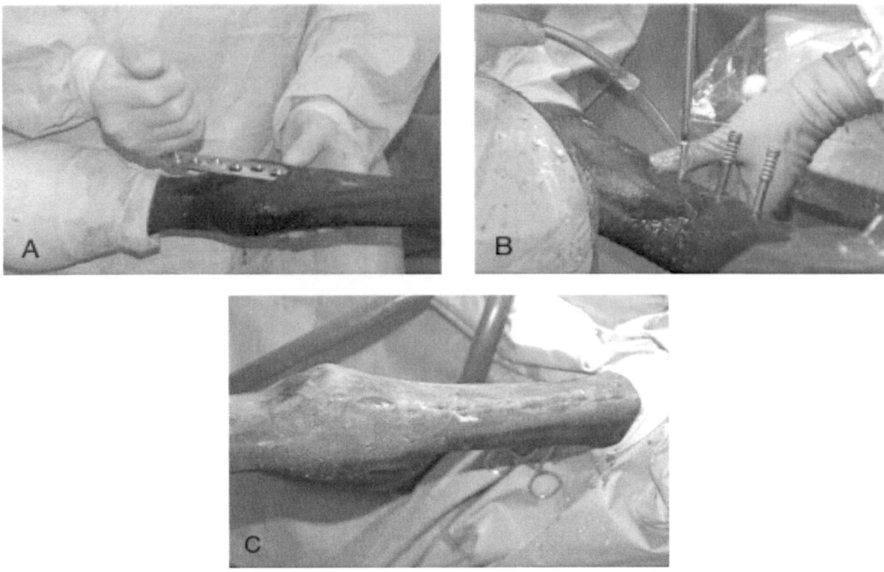

Pearls for performing MIPO

1. In order to achieve desirable outcome a few steps are critical: performing MIPO earlier the better as reduction and achievement of adequate length becomes difficult by indirect measures in older fractures; having accurate knowledge of surface anatomy as precise incisions and dissection is required; achievement of indirect closed reduction, avoiding exposure of the fracture and using special tools if required; making small incisions for implant insertion; use of plates preferably anatomically contoured with low bone contact as pure splints; elastic bridging of the fracture to formulate a relatively stable construct.

2. Attainment of intra-operative control over bone alignment and rotation of fragments can be cumbersome even to the experienced hands. Repeated use of intra-operative use of fluoroscopy exposes the surgeon to hazards of radiation. The following few signs can minimize the hassle while using fluoroscopy- Cortical step sign: the opposite bone ends have unlike cortical thickness; Diameter step sign: in presence of malrotation, diameters of proximal and distal ends appear to be of different dimensions; lesser trochanter sign: limb rotation leads to altered appearance of the lesser trochanter.

3. In order to minimize these hardships, development of better instruments and techniques is essential. These include better navigation tools, plates with newer types of instrumentation such as aiming devices, indirect reduction

tools such as clamps, pointed forceps such as collinear forceps, use of distracters, wire passers, modified t- handles and the f-tool.

4. Location of a plate hole without Fluoroscopy: Fluoroscopy is not required to locate a plate hole. Place identical plate over the subcutaneous plate and locate holes, then make small stab incisions. Use small hemostat to locate the holes.

Pitfalls

1. As the fracture site in not directly visualized, chances of mal-reduction exist which lead to malunion of fractures. Also, use of this technique limits its application in articular fractures where direct visualization of the joint surface is required. In such cases this technique is used in conjugation with standard open approaches.
2. As the procedure involves meticulous soft tissue handling and minimal exposure the procedure is considered technically demanding and a longer learning curve is required.
3. This technique is contraindicated in cases of open fractures with severe soft tissue loss with no coverage of exposed bone, as the fracture site is already exposed, associated neuro-vascular injuries as exploration of injured structure is required, pathological fractures and cases of non-union and delayed unions as these fractures require freshening of fracture margins with or without bone grafting.
4. As compared to conventional plating this procedure requires higher fluoroscopic control and thus exposes the surgeon to higher amount of radiation.
5. Implant removal can be tedious, as small incisions are required for plate insertion and often incisions are not in place of position of screws as swelling subsides. Large extensile incisions are then required for implant removal.

Imaging control during minimally invasive percutaneous osteosynthesis

High-quality preoperative diagnostic imaging is a key element in assessing fractures that may be amenable to minimally invasive approaches. Often a complete series of radiographs are adequate to define the fracture. Fractures that are difficult to interpret radiographically, such as a spiraling medial condylar fracture, are ideal candidates for three-dimensional imaging. Computed tomography (CT) is the gold standard to accurately evaluate fracture configurations. Currently, the availability of standing CT is limited, thus requiring horses to be placed under general anesthesia for placement into the CT gantry. As standing and robotic CT become more widely available, they may aid in preoperative planning.

Intraoperative imaging is vital for minimally invasive fracture repair. Fluoroscopy or radiography can be used to accurately facilitate implant placement. Fluoroscopy has the advantage of being under control of the surgeon and capable of producing real-time images. Precise use of fluoroscopy takes practice; obtaining perfectly positioned images intraoperatively can be challenging. It is vital that good-quality images are acquired in multiple planes when implant positioning is being evaluated. The surgeon should take the time to correctly position the fluoroscope and optimize the settings at the beginning of surgery. Intraoperative radiography can also be used during MIPO. Digital radiography is superior because of the rapid image acquisition. A sterile covering for the detector plate is used so that the surgeon can position it as desired. Personnel are required in the surgical suite to operate the x-ray generator. Another option that is becoming increasingly more attractive is intraoperative CT imaging, which is explained in the section CT-guided osteosynthesis later.

Basic principles of screw and plate insertion

Screw fixation through stab incisions is used commonly for many equine fractures including condylar fractures, fractures of the proximal phalanx, and carpal/tarsal slab fractures. Stab incisions should be made down to the periosteum and be large enough so that a drill guide can be inserted. For screw fixation in lag fashion, the glide hole is drilled first using intraoperative imaging to ensure the fracture line is entered with the drill. A Steinmann pin can then be inserted into the hole before the drill guide is removed to facilitate placement of the centering sleeve. The thread hole is drilled once the centering sleeve is fully inserted. A drill guide should be used to protect the surrounding soft tissue during tapping. Finally, the screw is inserted and tightened. Screw length and fracture compression are checked with the help of intraoperative imaging.

Locking compression plates are very useful for MIPO for reasons discussed earlier. Plates can be passed through small (2–4cm) incisions created at the proximal or distal end of the plate. Incisions should be made boldly through skin and subcutaneous tissues down to the periosteum. A plate-passing device is used to create a subcutaneous or sub-tendinous tunnel for the plate. A plate-passing device can be made from a broad DCP that is sharpened on one end and has a handle welded to the other end to facilitate pushing. A mallet can also be used to tap the plate passer into place. The plate-passing device can be contoured as needed using a large plate bender. Following creation of the tunnel, the plate can be slid in the appropriate position over the periosteum. The long drill guide threaded into one end of the LCP is useful for positioning the plate in the tunnel. Plate positioning is confirmed with intraoperative imaging and adjusted as needed. Once the plate is well positioned, two screws should be placed to ensure the plate does not move. A scalpel is then used to make stab incisions down to the periosteum over each plate hole. Care is taken to ensure that the LCP drill guides are fully and perpendicularly threaded into the plate to avoid oblique drilling, which can cause cross-threading of locking screws and inevitably would decrease the strength of the construct. Fracture lines cannot be directly visualized during MIPO; however, entering an occult fracture line is likely of minimal consequence when LCPs are used because of the increased stiffness of the construct.

MIPO is well suited for repair of lateral and medial condylar fractures that require screws inserted in lag fashion or plate fixation. Arthrodesis of the proximal interphalangeal joint and metacarpo-/metatarsophalangeal joints can also be accomplished using a minimally invasive approach. However, because the surgeon's ability to remove articular cartilage is limited, this technique should be reserved for cases of severe osteoarthritis with significant loss of articular cartilage and for horses with an intact suspensory apparatus. In minimally invasive approaches, partial cartilage débridement can be achieved by passing a large (5.5-mm) drill bit across the joint in several planes under fluoroscopic guidance. In the metacarpo-/metatarsophalangeal joint, cartilage débridement can be facilitated arthroscopically; however, arthroscopic débridement in the proximal interphalangeal joint is of limited value. Advantages of minimally invasive arthrodesis include less postoperative joint instability because of preservation of the collateral ligaments and joint capsule, and decreased SSI as a result of less soft tissue damage and smaller incisions.

Computer-assisted surgery

Computer-assisted surgery (CAS) allows the surgeon to accurately implement the preoperative plan and to implant screws at the desired location and at the correct angle relative to the fracture plane. CAS has been shown in numerous publications to improve accuracy in the placement of screws and other devices in humans.

CT-guided osteosynthesis

In contrast to CAS units, multislice helical CT units are available in an increasing number of equine referral centers. They do not only provide very exact information on the fracture configuration, but can also be used for intraoperative control of implant placement because of the very short image acquisition time. For this purpose, osteosynthesis is performed in the CT room or a portable CT unit (Ceretom) is used in the operation theater. Markers and an aiming device are positioned under CT guidance. Drilling is performed through the aiming device and the depth of the glide hole is also controlled with an intraoperative

CT scan. This technique allows very precise and minimally invasive placement of screws in lag fashion in anatomically difficult areas.

Osteosynthesis with bone graft

Most surgeons consider plate osteosynthesis with bone graft to be the gold standard for the treatment of atrophic humeral shaft nonunion. Plating with bone graft has been found superior to both ORIF alone and IMN fixation coupled with supplemental bone grafting. Either autogenous bone graft or demineralized bone matrix (DBM) are placed in the intramedullary canal and adjacent to the nonunion site. In a retrospective review of 45 patients with humeral shaft nonunion treated with ORIF and bone graft, found no difference in union rate following application of autogenous iliac crest bone graft (ICBG) versus DBM; however, it is reported that 44% of the ICBG group experienced donor site morbidity. Regardless of the origin, bone graft or other biologic stimulus can serve as a useful adjuvant to plate osteosynthesis.

Summary

Bone plating has been used as a method of fracture management for many years. Now, a trend toward the use of fracture fixation techniques which preserve the local fracture environment, known as biological osteosynthesis, has evolved. This trend has resulted in the development of a less traumatic method of bone plating referred to as minimally invasive plate osteosynthesis (MIPO), or percutaneous plating. During MIPO fracture stabilization, plates are inserted through short incisions and a communicating epiperiosteal tunnel. Typically, bone plates applied in this fashion have a bridging function. Promising outcomes have been reported in human patients undergoing MIPO fracture stabilization, and limited reports of the use of this technique in dogs and cats have yielded positive results as well. Careful case selection, pre-operative planning, and appropriate instrumentation are necessary when performing the technique. Rapid time to union, low complication rates and good return to function has been noted in human patients. Additional research is needed to define selection criteria and outline the definitive benefits of MIPO in dogs and other animals.

[Acknowledge for reproducing some text and photos from references: Richardson DW, Ortved KF: Minimally Invasive Fracture Repair, In Equine Surgery [5th Ed] e-book (Ed: JA Auer and JA Stick) Chapter: 13: P-229-232; Choudhari P and Baxi M: Minimally Invasive Plate Osteosynthesis: A Review Indian Journal of Orthopedic Surgery 2016, 2 (2); 194-198)].

23

Management of Comminuted Femoral Fracture by Different Internal Fixation Techniques

One of the most common fractures seen in the dog is comminutuion of femoral shaft, often related to sever impact from an automobile accident or jump from a height. Comminuted fractures are characterized by one or more butterfly fragments and assorted small bone chips that are especially difficult to align accurately and stabilize completely. These complicated fractures require sophisticated repair technique that are capable to resisting compression, bending and rotational forces. Rebuilding of comminuted fractures so that all fragments are accurately aligned and gaps are eliminated is often impossible. Non-union and osteomylitis have their highest incidence in these fractures. Over the years, several techniques are tried to treat comminuted femoral fractures with different success rate and there is a lack of uniform fracture presentation. Hence, it will be difficult to compare different techniques in a particular type of fracture. The present chapter is, therefore, undertaken to evaluate radiographically and histopathologically the efficacy of three different internal fixation techniques in the management of comminuted diaphyseal femoral fracture in dogs.

Clinically healthy, adult mongrel dogs (15) of either sex are randomly divided into groups A, B and C of 5 animals each. In all the animals unilateral comminuted diaphyseal femoral fractures is created under general (thiopental - 5%) anaesthesia and is immobilized with intramedullary (IM) pinning with cerclage wiring (Gr. A), neutralization bone plating (Gr. B) and neutralization bone plating with cerclage wiring (Gr. C) respectively. All the animals are administered broad spectrum antibiotic (streptopenicillin @ I g intramuscularly (dog/day) and analgesic (diclophenac sodium @ 1 mg/ kg b. wt. daily) and surgical wounds are dressed daily with povidone iodine and antibiotic cream for 5 days postoperatively.

Lateral and oblique views of radiography are taken at the operated site just after the operation and subsequently on days 15, 30, 45 and 60 postoperatively. The radiographs are observed for status of fracture reduction and fixation, progress of healing at the fracture site and complications like fixation failure, osteomylitis etc., if any, at different intervals. Bone sections are collected from the fracture site including adjacent normal bone after sacrificing the animals on day 60. The sections are decalcified using Goodling and Stewart's solution (80% distilled water, 15% formic acid and 5% formalin) and then processed routinely for H & E staining for histopathological studies

In group A, radiograph taken immediately after surgery, shows satisfactory reduction of fracture fragments with Steinmann pin and wires in proper position. However, slight fracture gap is present in almost all the animals. In one animal, the pin has crossed the distal cortex of the bone and it is later slightly retracted and left in position. However, fixation is maintained and no other complication is seen in this animal till the end of the observation period. In three animals, in which IM pin is placed almost occupying 80-90% of the medullary cavity. On day 0, the comminuted fractured fragments are either perfectly aligned or slightly deviated, and it is maintained throughout the observation. On days, 15 and 20 little periosteal callus formation is evident from the compression side of the bone. On day 45 periosteal reactions are more pronounced and extended to almost entire diaphysis. On day 60, the fracture gap is almost completely obliterated by bridging callus. Periosteal continuity is well developed at the fracture site. The union of the medullary cavity across the fracture line, however, is not completed and bone is in the process of remodeling. In two animals, signs of fixation failure like proximal pin migration with collapse of fixation due to slippage of cerclage wire into the defect site and rotational instability are found. By day 60, periosteal new bone formation is noticed but still fracture gap is evident in the proximal fragment. In this group, a relatively larger fracture gap is present might be due to inadequate compression provided by the cerclage wires on comminuted fragments. In three animals, in which IM pin occupied 70-90% of the medullary cavity facilitating three point fixation, fracture healing is normal without any complication and indicating adequate fixation. The bending strength of an IM pin is related to its diameter.

In group B, radiographs taken immediately after surgery shows excellent reduction and alignment of fracture fragments in all the animals. The fixation is maintained in all animals up to the end of observation period. On day 15, periosteal callus is not visible in any of the animals. On day 30, there is a decrease in fracture gap and increase in radiographic density near the fracture line. On day 45, marked periosteal reaction from both the fragments is noticed and fracture gap is obliterated by newly formed osseous tissue from the proximal fragment. On day 60, this periosteal reaction is further organized and fracture line almost completely obliterated and primary bony union took place. Periosteal continuity is well developed across the bony defect. However, medullary cavity across the defect is not completed. In this group, fracture healing is better than groups A and C. The fixation is maintained in all the animals till the end of the observation period. Radiographs at different intervals reveal primary bony union that took place in these animals without any type of complication. Bone plating with interfragmentary compression screw on the butterfly fragment (neutralization plating) provides the most stable reduction and is especially indicated in comminuted femoral diaphyseal fracture. Because such a plate neutralizes bending, rotational and shearing forces that are acting on the fracture, it is called neutralization plate. This neutralization plate provided rigid fixation in these animals and show better healing pattern than IM pinning with full cerclage wiring.

In all the animals of group C, the radiographs taken immediately after surgery, also shows good reduction and alignment of fracture fragments: However, in only three animals fixation is maintained throughout the observation period and fracture site shows normal course of fracture healing. On day 0, the comminuted fragment is well secured and perfectly aligned at the defect site by one inter-fragmentary screw and two full cerclage wires. On subsequent intervals, this reduction and alignment are maintained and very small amount of periosteal callus formation is noticed at the defect site. On day 60, the periosteal reaction is more pronounced and complete union of the fracture fragment is developed at the proximal site of the bone defect. In this stage, periosteal continuity and medullary cavity across the defect site is well advanced, though remodeling still continued. In one animal of this group (C), in which supracondylar fracture is noticed in the distal fragment on day 46, the course of fracture healing is normal up to day 45. However, fracture healing continued at the comminuted fracture site till day 60. In another animal of this group, osteopenia and concurrent osteomylitis is evident by day 45. In this animal, osteolytic area is noticed in the proximal fragment at the level of second screw, in between the plate and bone. On day 48, it is extends to whole length of the bone adjacent to the plate and by day 60, it is developed into an oblique fracture at the level of second proximal screw. In this group mixed results are seen. In three animals, where, fixation devices are remained intact up to the end of observation, radiographs of different intervals reveal little periosteal formation from both the fractured fragments, and bridging the fracture gap and primary bony union took place. The aim of compression treatment of fracture is to achieve primary bone union in which direct longitudinal reconstruction of the bone occurs without any radiologicaly visible periosteal or endosteal callus. In this group, occurrence of secondary supracondylar fracture and osteopenia followed by osteomylitis by 45 days in two animals may be because of the fact that compression plate excessively protected the bone from loading stresses. Since the bone requires some stress to stimulate healing, this "stress protection" can impede healing response may be evidenced by bone resorption in bone under the rigid plate (stress protection). Playful temperament of these animals may be another factor for supracondylar fracture. In these animals collapse of fixation devices may be due to loosing and slippage of cerclage wires into the fractured line, which is evident radiographically. Fixation of bone fragment by cerclage wiring has a deleterious effect on cortical blood circulation, subsequent bone healing and callus bridging. This is based on the concept that longitudinally directed afferent blood vessels are present in mature periosteum normally and during fracture fixation and that an encircling wire would 'strangle' this vascular supply.

The decalcified, H&E stained sections reveal fracture healing faster and completed in the endosteal region than the outer cortex. This is probably due to the availability of more bone cells, marrow cells, and rich blood supply available in the endosteal region. The defect site is identified by the presence of fibro-cartilaginous or chondro-osteoblastic activity at different areas within the fracture indicates that the fracture is healed by endochondral ossification

In group-A, shows a relatively larger fracture gap with more fibrocartilagenous activity and osseous tissue suggesting delayed union. The cartilage is seen arising from the cambium layer of the periosteum and is replaced by bone near the fracture site. Excessive chondroblastic and osteoblastic proliferation with abundant marrow tissue is seen along with osteoclastic activity. There are not fully developed bony trabeculae, which does not cross over the fracture line. This may be attributed to the fact that IM pins with full cerclage wiring unable to prevent the rotational motion at the defect site completely leading to delayed vascularisation and fracture healing. The initial fibrin scaffold formed during fracture repair, is disrupted if immobilization is inadequate. Cerclage appliances causes non-union when use along with IM pin fixation. This suggests that IM pinning with full cerclage wiring is unable to stabilize comminuted fracture and it further confirms the radiological findings.

In group B, fracture healing is better than groups A and C with the presence of numerous osteoid tissues at the defect site and new bone trabeculae is penetrating from both the ends towards the defect area. The newly formed osseous tissue (woven bone) near normal bony tissue is evidenced in an improper direction. Abundant bone marrow is also present with few osteoclasts near the bony spicules, thickened periosteum, and organized periosteal callus with vacuoles. Occasionally, soft callus is also seen at the endosteal side with a marked fibrocellular reaction. The marrow tissue is abundant and shows presence of osteoclastic activity. The fracture line is not clearly visible in most cases in this group. Blood vessels and mononuclear infiltration is also evident within the marrow tissue. Bone plating with interfragmentary compression screw on the butterfly fragment (neutralization plating) provides the most stable reduction and is especially indicated in comminuted femoral diaphyseal fracture as it neutralizes bending, rotational and shearing forces.

Group C animal's show mixed results. In two animals where fixation devices remain intact up to the end of observation, shows complete union with remodeling at the fracture sites. Presence of chondroblastic and osteoblastic activity and fibrocellular reaction near the periosteum suggest of fracture healing by cartilage formation. Endochondral ossification is seen with newly formed bone replacing the old cartilage. In between the cartilage at some areas newly formed blood vessels are also seen. Two animals of group C show non-union with fibrous tissue filling the fracture gap. Whereas, in one animal, osteopenia, compartmental syndrome and osteomylitis is evidenced by complete non-union at the defect area with the signs of bone necrosis and infection. These result suggest that full cerclage wiring with compression bone plating is not providing expected better results than bone plating when used alone (group B) in the management of comminuted femoral fracture. Fixation of bone fragment by cerclage wiring has deleterious effect on cortical blood circulation, subsequent bone healing, and callus bridging. The other inherent danger of cerclage wiring is that the wire may become loose, break or slippage into the fracture line causing fracture instability and produce an area of devitalisation by preventing the invasion of small blood vessels.

Summary

The study is conducted to evaluate and compare different internal fixation techniques for the management of comminuted femoral fracture in dogs. Identical comminuted femoral fracture is created under general anaesthesia in all the animals. The fracture is then immobilized by using intramedullary Steinmann pinning and cerclage wiring (group A), neutralization bone plating (group B) and neutralization bone plating with cerclage wiring (group C), respectively. In group A, fracture gap is present in almost all the animals. In three animals, fixation is maintained and good fracture healing is observed at the fracture site, whereas, in other two animals signs of fixation failure are observed. In group B, excellent reduction, alignment and fracture healing is observed in all the animals. In group C, radiographs shows mixed results. In three animals, normal course of fracture healing is observed while in other two animal's complications like fixation failure and osteomylitis are observed. Histopathologically, in group A, fibrocartilagenous activity is more pronounced and osseous tissue is present only at places suggests delayed union. In group B, primary bone union is took place with numerous osteoid tissues at the defect site and new bone trabeculae penetrating from both the cut ends towards the defect area. In-group C, observations shows mixed result. In two animals, where fixation devices remain intact up to the end of observation, complete union with remodeling at the fracture site is observed. Whereas in other three animals, non-union is evident with fibrous tissue filling the fracture gap.

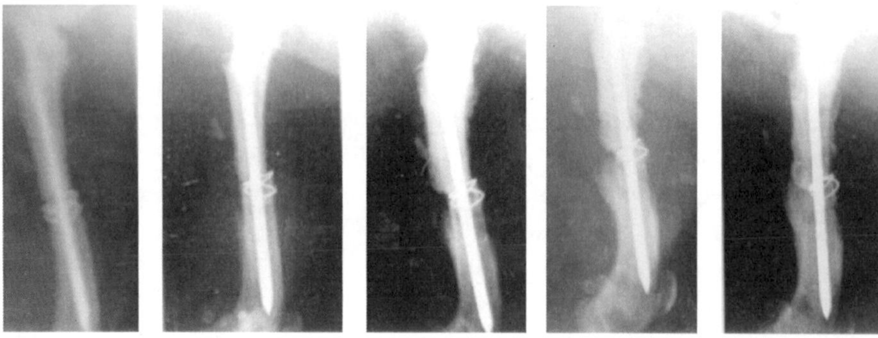

Day 0 15 30 45 60

Group A

Sequential radiograph of Group A at different interval

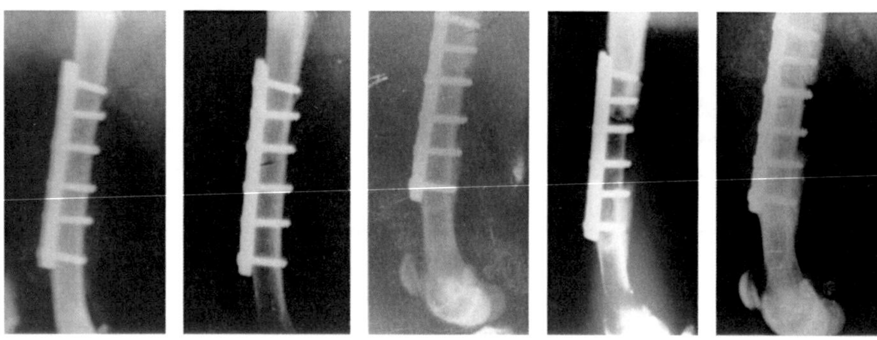

Day 0 15 30 45 60

Group B

Sequential radiograph of Group B at different interval

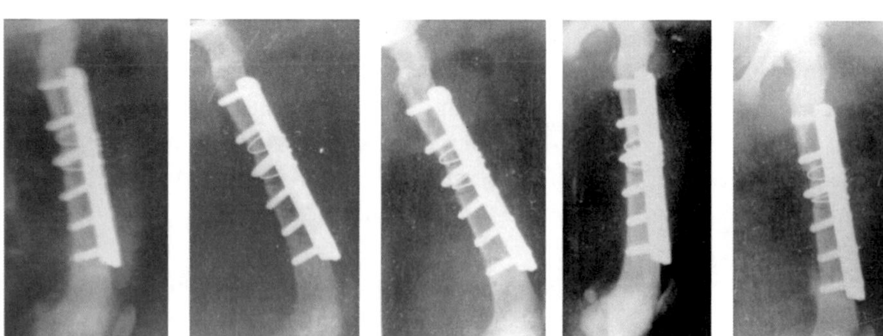

Day 0 15 30 45 60

Group C

Sequential radiograph of Group C at different interval

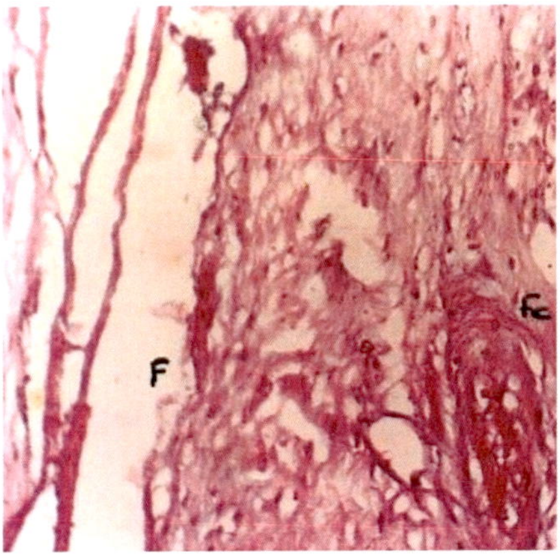

Photomicrograph of Gr A showing fibrocartilagenous tissue (Fc), fracture line (F)

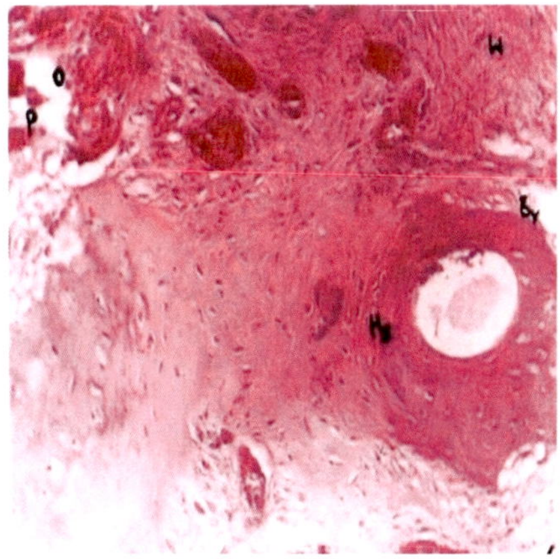

Photomicrograph of Gr B showing well developed Haversian system ((Hs), woven bone (w), new blood vessels (Bv), osteoclast (o) emerging from Periosteum (P)

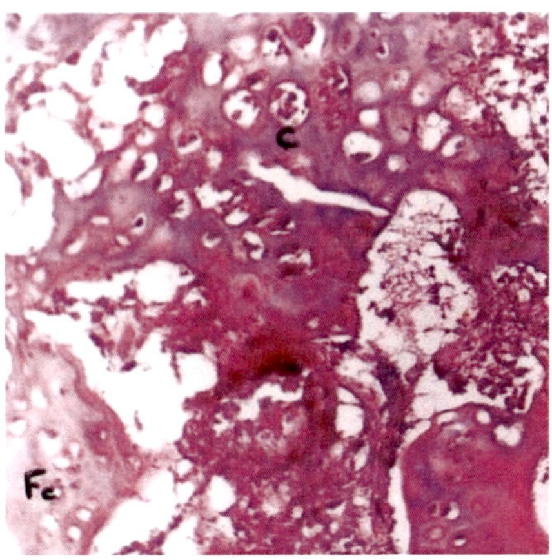

Photomicrograph of Gr C showing fibrocartilagenous tissue (Fc) with complete calcification of chondrocytes (C)

24

Comparative Study of Two Different Fixation Techniques in the Management of Tibial Metaphyseal Fracture

Among the epiphyseal and metaphyseal fractures of the different long bones, fractures of the proximal tibia are quite commonly reported and require special attention due to their proximity to the compound stifle joint. Hence, a standard fixation technique is needed to treat such fractures, which also permit early ambulation and preserve range of joint motion in addition to providing rigid fracture fixation. Different treatment modalities have been tried for fixation of metaphyseal fractures with variable results. No single method is appropriate in all circumstances and treatment indication often overlaps. Therefore, in this present chapter, an attempt has been made to evolve a standard technique for the management of such fractures.

Ten clinically healthy mongrel dogs are maintained under uniform housing, feeding and management conditions, acclimatized, immunized against rabies, dewormed and randomly divided into two groups (A and B) of five animals each. In all the animals unilateral metaphyseal fracture is created in the proximal tibia under general anaesthesia with intravenous thiopental sodium (5%). In the animals of group A, the fracture is immobilized with cross intramedullary Steinmann pinning with interfragmentary wiring, whereas, single intramedullary pinning along with interfragmentary wiring is performed in the animals of group B. Antiseptic dressing of the wounds is done daily for 1 week with povidone-iodine. All the animals are administered Enrofloxacin at the dose rate of 5 mg/ kg b.wt intramuscularly daily for 5 days posoperatively. Skin sutures are removed on day 10 post-operation

Intraoperative observation: Duration of surgery, extent of soft tissue manipulation and trauma, degree of easiness or technical difficulties in application of the implants, extent of fracture reduction and fixation and complication, if any, are recorded. *Clinical observations:* Respiration rate, pulse rate and rectal temperature of the animals are recorded before the start of the experiment and on day 1, 3, 7, 15, 20, 30, 45 and 60 post-operation. The operated limb of all the animals is observed for the following parameters on same days (a) wound condition, (b) degree of swelling, (c) postoperative pain on gentle pressing at fracture site, (d) weight bearing of the operated limb, (e) appetite, defecation, urination and general health of all the animals are observed throughout the period of observation. *Haemato-biochemical studies:* Venous blood samples from the saphenous vein are collected in all animals before and on same days and are analyzed for total and differential leukocyte count. The serum samples are analyzed for sodium and potassium, calcium, phosphorus and serum alkaline phosphatase. *Radiography:* Cranio-caudal and mediolateral views of the fracture site are taken just after surgery and subsequently on days 15, 30, 45 and 60 post-surgery. *Functional status of stifle joint:* The angle of extension and flexion of the stifle joint is recorded on days 15, 30, 45 and 60 after surgery using goniometer. *Biochemical and histopathological studies of callus:* After the animals are sacrificed on day 60 after surgery, 1 x 1 cm long bone sections are cut from the fracture site including normal bone on either side. After decalcification in Goodling and Stewart fluid, bone samples are divided into two pieces. One piece is subjected to the estimation of collagen and hydroxyproline and hexosamine. The hydroxyproline values are multiplied by factor 7.25 to get the collagen content. Second decalcified pieces are processed in routine manner for H&E staining.

All the animals of groups A and B are maintained good general health with normal appetite, defecation and urination throughout, the period of observation. In animals of group A, tissue manipulation is more than those in group B. The time taken to complete the surgical procedure is much longer in group A than B. In all animals, interfragmentary wiring is perfect with least gaping and very rigid fixation is achieved with excellent stability at the fracture site. The surgical wounds is healed by first intention in 8 days in all the animals except one animal of group A and two animals of group B, where suture dehiscence have occurred and the wound is healed by second intention in 12-15 days. Postoperatively, in all the animals of groups A and B, moderate inflammatory swelling is appeared on the operated limb on the 1st day which increased to severe by 7th day on which it is maximum in all the animals. The swelling is reduced slowly from day 8 and is mild on day 15. In the animals of group A, severe to moderate pain is recorded for the first 7 days then it is mild by the 15th day and is very mild to nil from day 21 till the end of observations. Wherein in group B, severe pain is shown by all the animals for the first 7 clays then it is moderate up to 30th day after which it is gradually declined to very mild to nil by the 50th day.

The technique of cross intramedullary pinning (group A) is selected keeping in view of its spring loaded tension effect and the three point fixation phenomenon. This technique is required maximum time for fixation, tissue manipulation is relatively more and its application posed comparatively more difficulty than the single pinning. This is mainly because 30° angle of normograde insertion of the pins, at the point of insertion of collateral ligaments at the tibial plateau, is quite critical to avoid penetration of the opposite cortex. However, once properly it's reduced and fixed, the fixation achieved in all the cases is more stable than that of single intramedullary fixation.

In all the animals, the signs of inflammation are appeared on first postoperative day and it is reached maximum on day 7. The inflammation is relatively more in group B and it is persisted for longer duration due to less rigid fixation of fractured fragments. Animals of all the groups are started bearing weight from the very first postoperative day indicates good fracture fixation. However, weight bearing is slightly reduced during the period of intense inflammation. Weight bearing in group B is relatively slow and it is remained significantly less as compare to group A. This may be attributed to rigid fixation and early fracture healing in animals of group A.

Significant increase in TLC, neutrophil count and concomitant decrease in lymphocyte count is observed during early postoperative period in all animals. The values are returned towards normal after 10 days indicating resolution of inflammation and stress. Leukocytosis due to neutrophilia has been reported by corticosteroid release in states of stress, pain, trauma and surgical manipulation. The serum sodium, potassium and calcium levels are decreased in the immediate postoperative period reaching normal or slightly above the base value at the end of observation period. It's also reported that just after surgery or severe trauma, the very act of getting well may result in secondary hypokalemia. Whereas, serum phosphorus level is increased immediately after fracture fixation in both the groups and it is remained above normal level till the end of observation period. There is an increase in the serum alkaline phosphatase (AP) actively immediately after surgery in both the groups. The enzyme activity is further increased from day 10 onwards upto the end of observation period. The increased AP activity may be suggestive of an increase in the osteoblastic/chondroblastic proliferation during fracture repair.

Radiographs are taken immediately after surgery show satisfactory reduction and alignment of fracture fragments in animals of both groups. However, little fracture gap is invariably present in all the animals and the gap is relatively more at the cranial edge of fracture site than the caudal edge. This can be attributed to the greater tensile forces generated by the extensors of the stifle like quadriceps femoris muscle acting on the cranial edge. The interfragmentary wire is applied at this edge though provides good apposition, can not neutralize the tensile forces completely. The gap is very negligible at the caudal edge due to compressive forces. The animals of group A shows early signs of faster fracture healing than that of group B. This may be attributed to relatively more rigid fracture fixation achieved by cross pinning as compared to single pinning. Cross pinning provides a three point fixation under a spring loaded tension and thus helps to neutralize most of the fracture forces. Whereas, in animals of group B, single pin can not resist the fracture forces completely especially the rotational force. In three animals of this group (B), fracture is failed to heal as evidence by splintering of bone and progressive increase of gap at the fracture site, without any sign of organized bridging callus at the end of observation period particularly at the cranial edge of fracture may be due to the extensive tensile forces and less rigid fixation.

In both the groups, the extension angle is more affected than the flexion one. Moreover, the recovery of the flexion angle is slightly faster than extension angles. Probably, this is due to the fact that the extensors of the stifle joint are more closely associated with the joint than the flexors. Both the angles are near normal by day 45 in group A but in group B, improvement is not observed even by day 60 probably due to delayed healing and occurrence of fibrous union at fracture site.

The mean collagen, hydroxyproline and hexosamine content of the callus at the fracture site in group A are 13.18±1.59 g %, 1.81±0.22 g % and 550.48±23.26 mg %, respectively. Whereas, in group B, it is 9.81±0.84 g %, 1.33±0.11 g % and 305.80±29.14 mg %, respectively and these values are significantly lower than that of group A. This is may be due to rigid fixation achieved in group A, leading to normal course of fracture healing.

In all the animals, fracture healing is faster and complete in the inner cancellous bone than the outer cortex. This is probably due to the availability of more bone cells, marrow cells and rich blood supply in the cancellous bone. The healing is better in group A with the presence of different resorption cavities, osteoid tissue at the fracture site and new bone trabeculae penetrating from both the end. In most of the animals of this group, chondro-osteoblastic activity is noted at places. In contrast, in group B, fibrocartilagenous activity is more pronounced and osseous tissue is present only at places, where fracture gap is distinct, suggesting delayed healing. In two animals of group B, fracture site is completely filled with fibrous tissue, with little or no signs of ossification, indicating non-union, may be due to the fact that single pinning is unable to prevent the motion/forces at the fracture site completely.

Summary

In ten clinically healthy dogs, divided equally in two groups, unilateral metaphyseal transverse fracture is created, in the proximal tibia under general anaesthesia.

The fracture is then immobilized with cross intramedullary pin fixation in group A and single pin fixation in group B, along with interfragmentary wiring. These two different fixation techniques are compared based on intra-operative, clinical, radiological, haemato-biochemical, histopathological and functional studies of the test stifle joints. Clinical observations and radiographs taken on different intervals reveal excellent fracture fixation and faster healing in group A. Extension and flexion angles of stifle joints is improved steadily and reached near normal by day 60 in group A, however, in group B, it is remained significantly low till the end of observation period. Organic components of the callus at fracture site is markedly increased in groups A than B. Histopathological examination reveal complete healing along with remodeling at the fracture site of animals of group A, whereas delayed/non-union is observed in animals of group B. It may be conclude that single pinning is not enough to immobilize the fracture site for proper healing.

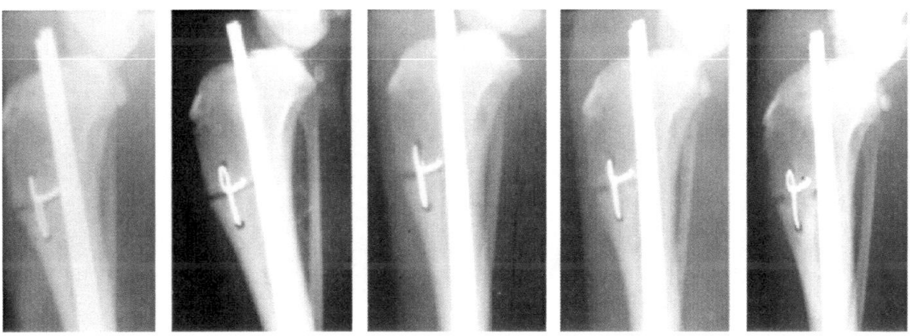

Day 0 15 30 45 60

Group A

Lateral radiographs of Gr A at different intervals showing normal fracture healing

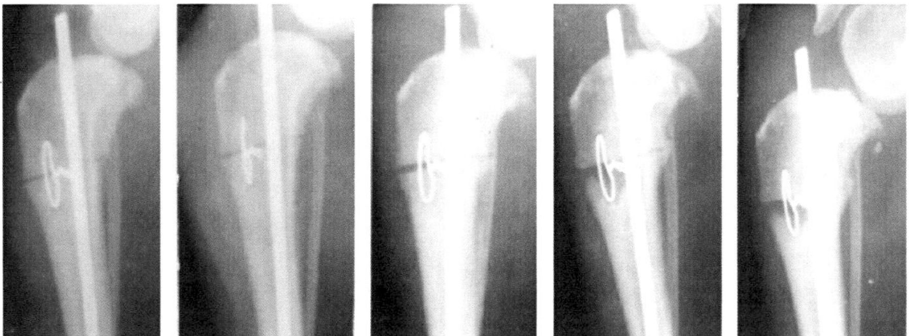

Day 0 15 30 45 60

Group B

Lateral radiographs of Gr B at different intervals showing delayed fracture healing

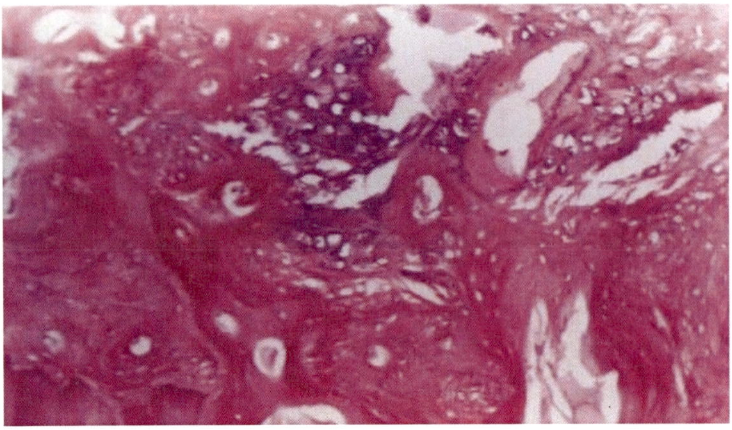

Histopathological section of Gr A showing complete bone union between the cortices at the fracture site with different resorption cavities

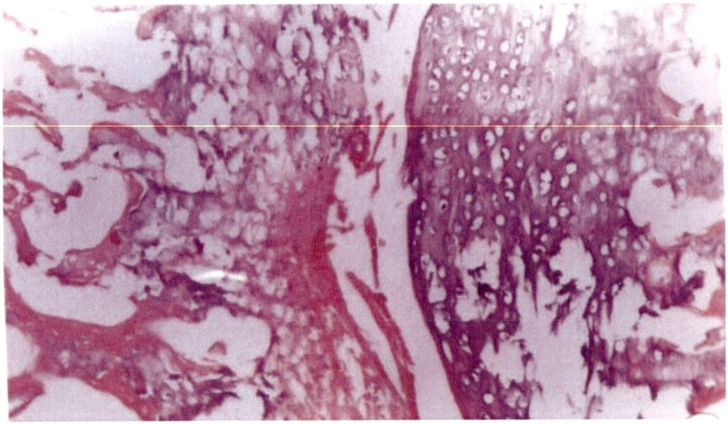

Histopathological section of Gr B depicting the presence of fibrochondroblastic activity at the fracture site

25

Endocrine Responses on Fracture Healing

Healing of fracture is a consequence of cellular behaviour in a living organism and as such can be modified by almost all endogenous or exogenous factors that affect the metabolic function of cells. These factors include transforming growth factor-β, bone morphogenetic proteins, hormones, vitamins and prostaglandins. The anabolic hormones like thyroxine and insulin accelerate bone formation in experimental animal. Hence, this chapter is proposed to evaluate the effects of thyroxine and insulin therapy in the management of tibial metaphyseal fracture in dogs.

Clinically healthy, adult mongrel dogs (15) of either sex are randomly divided into three groups (A_1, A_2 and A_3) of 5 animals each. In all the animals, unilateral metaphyseal fracture is created in the proximal tibia and then immobilized with cross intramedullary pinning with interfragmentary wiring under thiopental (5%) anaesthesia. All the animals are administered broad-spectrum antibiotic (streptopenicillin @ 1 g intramuscularly /dog/day) and analgesic (5 ml I/M daily) and surgical wounds are dressed daily with Povidone iodine and Lorexane cream for 5 days postoperatively. In group A_2, thyroxine sodium tablets (0.013 mg/kg body wt, orally) and in A_3, insulin injection (2 IU/kg body wt, s/c) are given on day 7 and then on every alternate day till day 45 postoperatively, at one specific time. Group A_1, is remained as control, in this group no hormonal therapy is given.

Craniocaudal and mediolateral radiographs of the fracture site are taken immediately after surgery and subsequent on days 15, 30, 45 and 60 postoperatively. The radiographs are observed for status of fracture reduction and fixation immediately after surgery and subsequently for the maintenance of reduction, progress of fracture healing and complication, if any, at different intervals.

Oxytetracycline @ 25 mg/kg b wt is injected intramuscularly on days 47, 48 and after 6 days intervals on days 55 and 56 (2-6-2) postoperatively for double toning of the bone. After euthanizing the animals at the end of observation period (day 60), 2-3 mm sections are cut from the fracture site longitudinally including the surrounding old bone. Undecalcified ground sections are then utilized for evaluating the nature and extent of the new bone growth at the fracture site by observing under ultra-violet light (Leitz Otrthoplan Universal Widefield microscope with incidental light). Bone sections are collected from the fracture site after euthanizing the animals at day 60 postoperatively. The bone sections are preserved in 10% formalin for 7 to 10 days and then they are decalcified using Goodling and Stewart's solution. Completely decalcified bone sections are then processed routinely for H&E staining. These sections are observed for status of healing at fracture healing.

Radiographs taken immediately after surgery depict satisfactory reduction and alignment of fracture fragments in all animals. However, little fracture gap is invariably present in all the animals and the gap is relatively more at the cranial edge of fracture site compare to caudal edge. This can be attributed to the greater tensile forces generated by the extensors of the stifle like quadriceps femoris muscle acting on the cranial edge. The inter-fragmentary wire is applied at this edge though provides good apposition, cannot neutralize the tensile forces completely. At the caudal edge the gap is very negligible due to compressive forces. Radiographs taken at subsequent intervals show maintenance of implant and fixation in all the animals indicates proper reduction and fixation of fracture.

In the group A_1 (control), radiographs taken at different intervals does not show any change in the position of either pins, wires or fracture fragments except in one animal where one pin is seen to have crossed the distal cortex of the tibia and in another animal, splintered fracture is occurred at day 15. However, fixation is maintained and no other complication is seen till the end of the observation period. Periosteal proliferation at the site of bone defect is not visible in the radiographs on day 15, though the periphery of the defect becomes smooth and uniform. In the day 30-radiograph, there are indication of new bone proliferation from the periphery. In almost all the animal's periosteal and intercortical callus is seen adjacent to the area of bone defect which is obliterated the defect site. On day 45, fracture gap is further reduced by bridging callus and it becomes more organized. Though fracture healing is perfect at the caudal border of the defect, little gap is still present at the cranial border and remodelling process is still continued in all animals on day 60.

Radiographs taken at different intervals in thyroxine-treated group (A_2) is revealed excellent bone healing in almost all the animals of this group without any complication, except in one animal, wherein splintered fracture is occurred, as seen in one animal of group A_1. At day 15, the periphery of the fracture line becomes uniform and smooth. In all the animals there is presence of new bone formation, as evident by periosteal and intercortical new bone formation adjacent to the area of bone defect. Fracture line is appeared irregular and hazy. On day 30, periosteal bridging callus become organized and obliterated the fracture gap. Further organization of bridging callus and obliteration of fracture gap is seen in all animals on day 45. Bone defect at caudal border is completely obliterated by new bone. On day 60, there is complete obliteration of fracture gap even at the cranial border by organized new bone and earlier remodeling of periosteal callus started in all animals. There are no radiographic signs of either delayed/non/malunion or bone infection in any animal, at any interval. In general, the healing is best in this group of animals when compare among 3 groups.

The fixation is maintained in all the animals of A_3 group up to the end of the observation period and all the animals' shows normal course of fracture healing, without any complication. On day 15, though the periosteal callus is not markedly pronounced as in group A_2, the radiographs reveal an increased density at the fracture site. On day 30, periosteal callus is distinguishable. The fracture gap is reduced substantially in all animals. On day 45, radiographs reveal further reduction in the fracture gap and the bridging callus becomes more organized. On day 60, excellent healing at the fracture site is observed with the presence of periosteal and intercortical callus and the cortical continuity is appeared to be normal. In general, the healing is better in A_3 than that in group A_1, but the amount of new bone formation is lesser than that of group $A_{2.}$

In the animals of A_1, the newly formed bony tissue is present almost completely in between the fracture fragments indicating bony union at the fracture site and suggested normal healing process. Also, the continuity the cortex across the fracture line is restored. Under u-v light, golden yellow fluorescence is seen within the fracture site suggesting the presence of good amount of newly formed osteoid tissue. Relatively less fluorescence is seen at places suggesting that the fracture gap is not completely filled with new osteoid tissue. A gap is seen at the cranial edge with fluorescent new bone which can be attributed to the greater tensile stress acting on the edge. In the animals of A_2 wherein thyroxine hormone is given, excellent fracture healing is observed as evident by complete bony union at the fracture fragment and restoration of continuity of the cortex. Under fluorescent microscope, the fracture line is visualized as a thin line of golden yellow fluorescence in between the fragments, whereas, the host bone evinced dark sea green homogenous colour. Within this new osteoid tissue, which completely filled the fracture gap, crossing over of numerous new bone trabeculae is evident. In group A_3, wherein insulin therapy is given, almost complete union at the fracture site is evident by the presence of newly formed less dense osteoid tissue and the continuity of the cortex at the fracture site. Under u-v light, good amount of golden yellow fluorescence at site of fracture is suggesting the penetration and crossing over of newly formed bony trabeculae from both the ends. Numerous resorption cavities lined with fluorescence yellow new bone are indicative of remodelling phenomenon. However, in this group, new bone formation and fracture healing is relatively slower than group A_2, but faster than group A_1.

In the group A_1 almost complete union at the fracture site is evident histopathologically, though the fracture gap with the newly formed bone can be well distinguished from that of normal bone. The newly formed osseous tissue near normal bony tissue is evinced in an improper direction. There are evidences of the presence of bony trabeculae which however, are not fully developed and do not cross over the fracture line. In group A_2 complete union and excellent remodelling at the fracture site are observed. Newly formed osteoid tissue is seen lying adjacent to the old bone, suggesting that it can be probable fracture line. Abundant bony spicules are present at the fracture site, with the presence of osteoclasts around them. At places, newly laid down trabecular bone is seen crossing over on either ends at the periosteal and endosteal sides. The trabeculae are consolidated and thickened. The periosteum is also thickened and infiltrated with mononuclear cells. There is abundance of marrow cells with presence of osteoclastic activity. Resorption cavities and replacement by newly formed osseous tissue is also the characteristic feature of this group. In group A_3 complete union at the fracture line with excellent remodelling is observed. The fracture gap is filled with near normal osteoid tissue which is evinced by its improper direction. In most of the cases the fracture line is almost invisible. Numerous newly formed thickened bony trabeculae penetrating are under progress, at both the ends.

The animals of hormone-treated groups (A_2 and A_3) show early signs of fracture healing compare to control group (A_1). This indicates that exogenous administration of thyroxine and insulin has helped to enhance osteogenesis and fracture healing. The thyroid hormone accelerates bone healing by influencing different phases of endochondral ossification. Thyroxine produces reduction in the immature tissue components forming more of cartilaginous tissue in the callus initially followed by quicker transformation into bone leading to early remodeling which is also found in this study. Insulin is associated with increased collagen synthesis, stimulation of osteoblastic activity and proper utilization of glucose and hexophosphates. These hexophosphates are important source of phosphorus for calcification of bone matrix. It's also reported that early replacement of the hypertrophied cartilage tissue by mature collagen fibers and new bone formation in the insulin-treated fracture cases. Similar observation is also observed in this study. These findings suggest that thyroxine and insulin administration helps to enhance the process of fracture healing, however, thyroxine have little edge over the insulin.

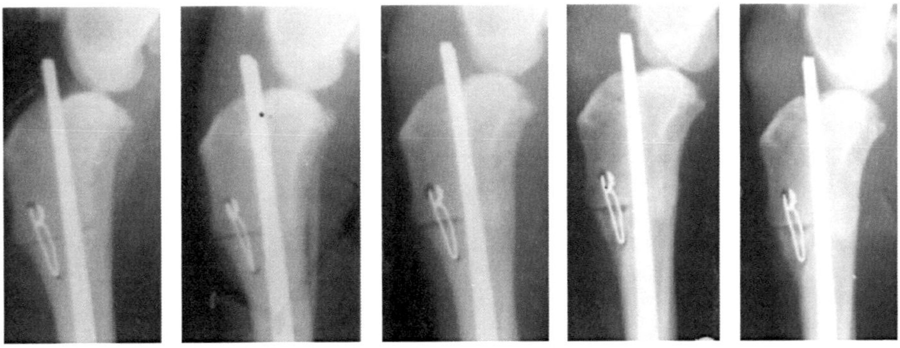

Lateral radiographs of A_1 in different intervals showing normal fracture healing

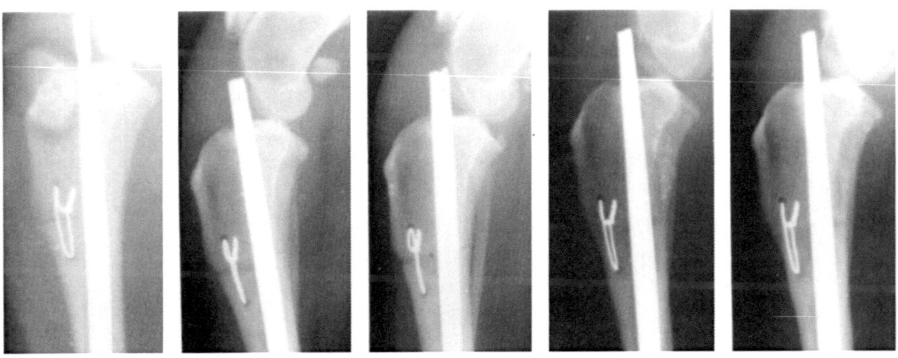

Lateral radiographs of A_2 showing accelerated course of fracture healing

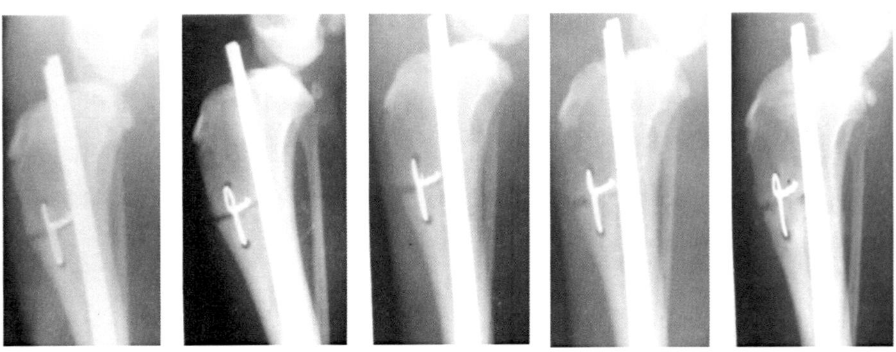

Lateral radiographs of A_3 showing accelerated course of fracture healing

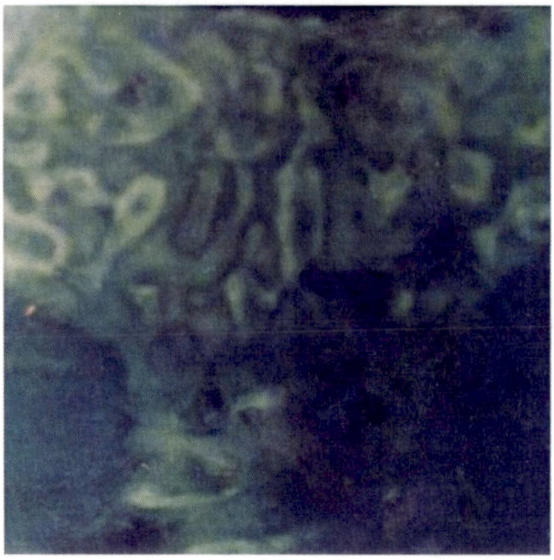

A1

A2

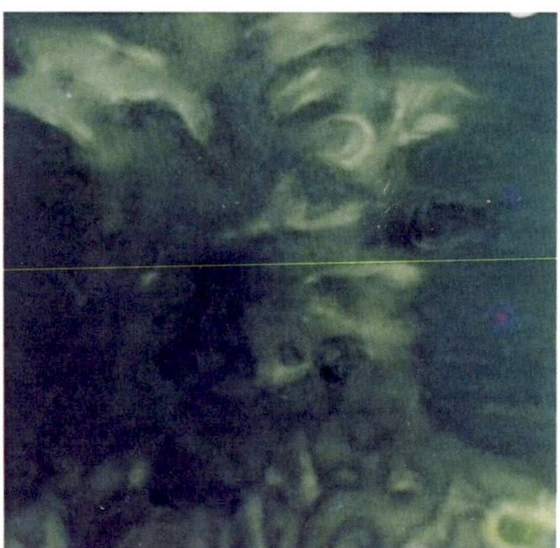

A3

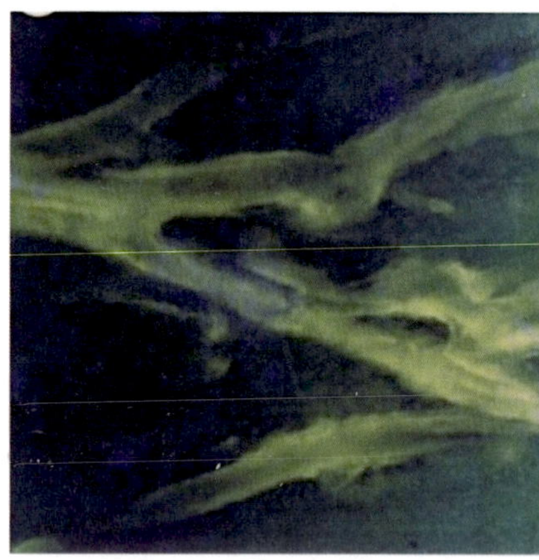

Fluorochrome labeled photomicrograph showing crossing over and union of bone trabeculae from either ends, in animals of gr A_1, A_2 and A_3. Photomicrograph shows formation of bone along the tract of the passage of cross pins (Right below)

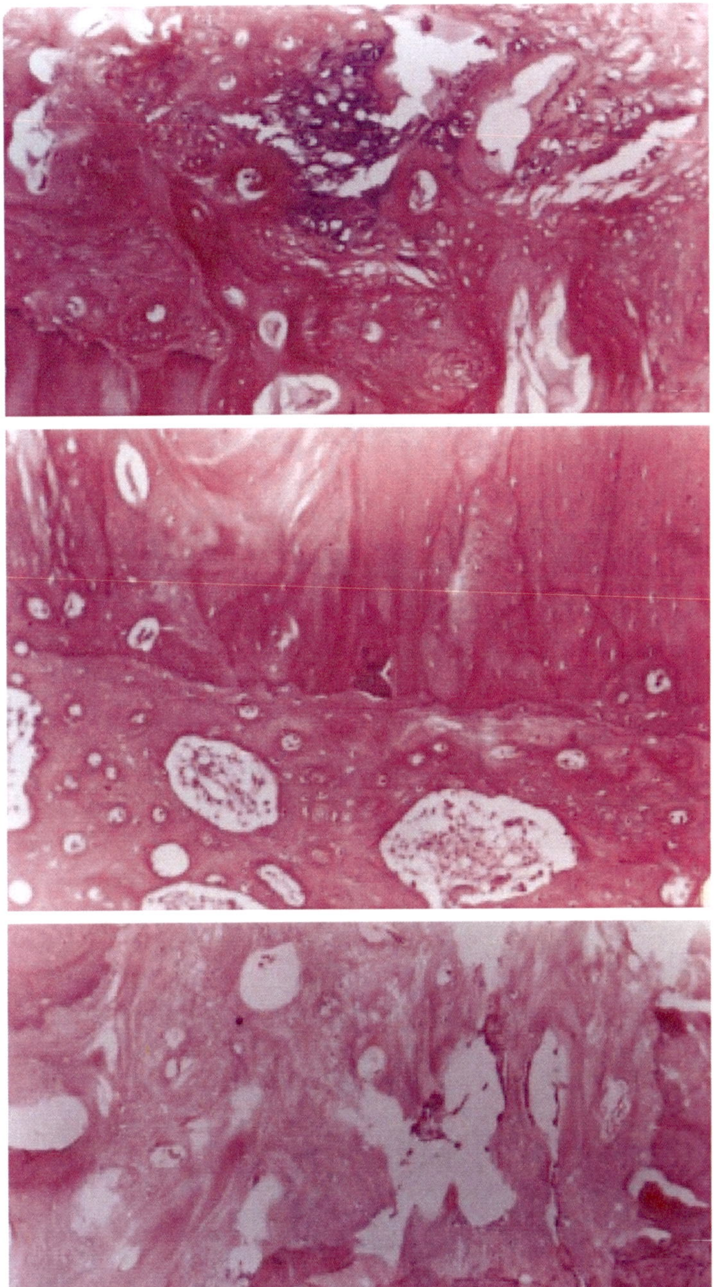

Histopathological photomicrographs of Gr A_1 (above), A_2 (middle) and A_3 (blow):

A_1- presence of chondroblastic activity at the fracture site, A_2- complete union between the cortices at the fracture site with resorption cavities, A_3- complete union between the cortices at the fracture site with presence of resorption cavities.

26

Evaluation of Composite Bone Grafts and Ceramic Implants in Fracture Healing

Bone grafting is important in the treatment of non-unions, delayed unions, joint arthrodesis, and filling of bone cavities and the replacement of bone lost due to cyst, trauma or tumours. An ideal bone graft or substitutes should be a material that is biologically inert or bioadaptable, a source of osteogenesis, having osteoconductive property, act as a mechanical support, readily available, easily adaptable to the site in term of shape and size and replaceable by the host bone. The beneficial effect upon osteogenesis of impregnating allograft bone with autogenous red marrow or autogenous citrated plasma has been the subject of current research. It is claimed that autogenous fresh bone marrow and citrated plasma are two useful contributors for osteogenesis. Biodegradable and bioadaptable ceramics like ALCAP (Aluminum-calcium-phosphorus oxide) seem to be the most promising biomaterials for correcting bone defects, since these ceramics encourage self-replacement by endogenous bone. However, use of bone grafts and ceramics with autogenous bone marrow or citrated plasma has not been investigated so far. This chapter is undertaken to evaluate bone marrow and plasma impregnated bone allografts and ceramic implants in fracture healing in goats.

Study protocol & grouping

Eighteen, locally available nondescript male/female goats of 12-14 months of age are selected on the basis of their clinical status and are kept under observation for a period of one month before start of experiment. The animals are housed in their shed and are allowed to graze in the surrounding pasture. They are kept on standard feed and fodder throughout the experiment. The animals are randomly divided into six groups consisting of three animals each. Two trials of a similar experiment are performed in each goat. A total of 36 trials are conducted to evaluate different bone grafts and ceramic implants (six trials for each group with three animals each): Group A, frozen decalcified allogenic bone grafts; Group B, frozen decalcified allograft impregnated with autogenous bone marrow (composite bone graft); Group C, frozen decalcified allograft impregnated with autogenous citrated plasma (composite bone graft); Group D, ALCAP ceramics; Group E, ALCAP ceramics impregnated with autogenous bone marrow; Group F, ALCAP ceramics impregnated with autogenous citrated plasma.

Collection and preparation of different types of grafts/implants

One clinically healthy goat of the same age group is euthanized. Bone grafts of desired size are then collected from the diaphysis of the metacarpal and metatarsal, taking all possible aseptic precautions. The grafts are then washed thoroughly with sterile normal saline. The decalcified allogenic bone grafts are prepared by keeping these grafts in 0.6 N HCL at 4^0 C till they are completely decalcified. Once a pin can penetrate easily through the graft, it is considered completely decalcified. These grafts are washed thoroughly in sterile distilled water and are kept in sterile normal saline solution overnight to ensure complete removal of acid. Afterwards, these decalcified grafts are kept in liquid nitrogen at -196^0C for a period of 7 days. The grafts are transferred to sterile normal saline solution containing antibiotic before their use in group A. Composite grafts are prepared by impregnating decalcified bone grafts with autogenous bone marrow and autogenous citrated plasma in groups B and C respectively. ALCAP ceramic implants is used as block, are cut to the appropriate size and shape and are sterilized in a hot-air oven at 100^0C for one hour. The implants are then allowed to cool down before their implantation in group D. The sterilized ceramic implants are thoroughly impregnated with autogenous bone marrow and autogenous citrated plasma as per methods used for composite bone grafts, before their implantation in groups E and F.

Surgical procedure

All types of grafts/implants are transplanted into a standard segmental bone defect (1 x3 cm^2) created in the proximal third of the medial aspect of the tibia using an electric bone saw under general anaesthesia. The grafts/implants are secured in position by suturing surrounding soft tissues. Both tibia of an animal are used for the experiment and only one type of graft/implant is placed into both defects. Routine postoperative care is given for a period of 5 days to each animal.

Evaluating parameter

(a) Radiology

The healing process, osteoinductive property and fate of different types of bone grafts and ceramic implants are evaluated and compared on the basis of radiological observation (lateral radiographs) made on days 0, 30, 60 and 90 postoperatively.

(b) Osteomedulography

On day 60 postoperatively, animals are preanesthetized with Xylazine 10 min before using local anesthetics at the site of drilling. One small hole is drilled at distal metaphysis using a 14 gauge bone marrow biopsy needle into the medullary cavity of the test bone (tibia). The correct positioning of the needle in the medullary cavity is assessed by the presence of blood on aspiration in a glass syringe attached to the needle. The heparinized normal saline is then infused for flushing the medullary cavity. Ten milliliters of sodium iothalamate (Conray-420) as contrast media is rapidly injected into the medullary cavity and lateral radiographs are taken during and immediately after injection. Osteomedullogram are observed for the venous drainage at the grafted/implanted area.

(c) Angiography

Tibial angiography is performed after euthanizing the animal with an over dosage of thiopental sodium at day 90 postoperatively. The tibial arteries are exteriorized and catheterized using polyethylene catheters. The arterial system of the limb is flushed thoroughly using heparinized normal saline till the clean saline started coming out of the veins. Freshly prepared 20% lead oxide soap suspension is then infused into the arterial system of the test limb. The vessels from which lead oxide suspension is flowing out are ligated. The tibial arteries are also ligated after complete infusion of the limb. The intact limbs are then washed thoroughly in running water to remove lead oxide from the limb surface and lateral radiographs are taken. Lateral radiographs are taken immediately after removal of all soft tissue of the test bone (tibia). The angiograms are observed for vascular response of the host bone and surrounding soft tissue in the grafted area and vascularisation of the graft/implant.

(d) Triple fluorochrome labeling

Triple fluorochrome, viz, tetracycline in the form of oxytetracycline (50mg/kg, i/m), hematoporphyrine hydrochloride (100mg/kg/IV) and DSAF solution (20mg/kg/IV) are used for labeling newly formed bones and resorbed bones. This triple fluorochrome are given at 24 hr interval before euthanizing the animals at day 90 postoperatively and grafted/ impaired segments of the bone are collected. A transverse section of 2-3 mm thickness including the grafted /implanted area is cut with the help of heck-saw. The sections are then ground to 20 μ thickness using different grade sand papers. Final grinding is done over the bone under moderate pressure using slow circular motion. The ground undecalcified sections are observed under ultraviolet light (Leitz Othoplan Universal) Widefield microscope with incidental light.

(e) Histopathological examination

The grafted bone/ ceramic implants are collected from the animals sacrificed at day 90 postoperatively. The sections are cut (3-4 mm thick) using hack saw including both normal and grafted area of the bone. The bone pieces are washed thoroughly with normal saline and fixed in 10% formalin for 7 days. Subsequently bones are decalcified in Goodling and Stewart's solution. The decalcified tissues are processed in a routine manner and stained with hematoxylin and eosin.

Results

In the lateral radiographs taken immediately after grafting, the grafted area is appeared as dark rectangular space. The decalcified bone grafts are not visualized suggesting their complete decalcification. At day 30, the periphery of the defect becomes uniform and there are indications of new bone proliferation from the periphery. At day 60, the marked decrease in the size of bone defect is seen due to organized periosteal proliferation. The defect is appeared elliptical in shape. The defects are almost completely filled with newly formed osseous tissue in three limbs, at day 90, whereas, in other three limbs, small area infiltrated with dense osseous tissue is seen. The periosteal new bone is in the process of remodelling. The bone marrow impregnated decalcified bone grafts cannot be visualized and the defect is appeared as dark rectangular space at day 0 radiography. Ingrowths of newly formed bone from the periphery of the bone defect as evidenced by the decrease in the size of defect at day 30. At day 60, a further decrease in the size of bone defect is observed and the grafted are is appeared cloudy due to the presence of osseous tissue. The periosteal new bone is well organized along the surface of the bone. The grafted area is almost completely filled with newly formed bone in all the six limbs. However, a small oval shaped area at the center of the defect is relatively less dense at day 90. The periosteal callus is in the process of remodelling. Plasma impregnated decalcified bone grafts are radiolucent and therefore cannot be visualized in the radiographs taken immediately after grafting. There is induction of new bone ingrowths into the defect from the periphery. However, there is no evidence of osteoid tissue formation within the graft at day 30. At day 60, marked reduction in the size of bone defect is seen due to new bone formation from its periphery obliterating the defect. At this stage, there is no evidence of graft contribution towards new bone formation. At day 90, in four out of six limbs, the area is completely obliterated with newly formed bone. However, in two limbs, small defect is still visible suggesting incomplete resorption and replacement of the graft. In these limbs unevenly distributed are of calcification/osteoid tissue is seen in the grafted area. In group D, lateral radiographs are revealed that the ceramic implants are in position filling almost entire gap/defect at day 0. The radiographic density of the implant is comparatively more than the host bone. There is no change in the size or density of implants at day 30. At day 60, newly formed bone is seen encapsulating the ceramic implants. There is no change in the size of implant. However, there is slight decrease in density of implant particularly at their periphery. At day 90, the radiographs are demonstrated that the activity of new bone formation ceased, newly formed bone is remodeled, and the implants are encapsulated. In group E, the bone marrow impregnated ceramic implants are well secured in position at day 0. At day 30, the edge of the implants becomes smooth and new bone approaching the implant from the periphery is seen. No significant change in size, shape or density of the implant is observed. The further growth of osseous tissue is seen into the defect at day 60. Resorption of implants is seen in some animals. At day 90, further resorption of implant and replacement by the new bone is seen. Periosteal new bone is organized and is in process of remodelling. In group F, the lateral radiographs taken immediately after implantation reveals that the implants are well secured in position. Although the signs of periosteal new bone formation on the surface of bone and around the implant are seen but no marked change in the size/shape or density of the implant is observed at day 30. At day 60, the newly formed bone is encapsulated the implant and the periosteal new bone is organized. Implants are present in all the limbs without significant change on size and shape at day 90. Slight reduction in their density, however, is suggested that the implants are in process of resorption. There is no evidence of implants contribution towards new bone formation at any stage of observation.

The osteomedullogram taken at day 60 shows flow of contrast material across the grafted site visualizing proximal sinusoidal network and extraosseous veins draining the contrast material from the medullary cavity of the experimental bone (tibia). Pooling or stagnation of contrast material is seen at the grafted site is greater in animals where bone marrow impregnated frozen decalcified allograft (Group B) is used followed by plasma coated frozen decalcified allograft (Group C) and frozen decalcified allografts (Group A). Stagnation of contrast material is suggested partial blockage of the medullary cavity due to endosteal new bone formation. All groups (D, E and F) where ALCAP ceramics alone or in combination with bone marrow and citrated plasma are used show mild to moderate stagnation of contrast media distal to the implanted site, cross-over circulation across the implanted site, proximal sinusoidal network and extra-osseous veins. There is no evidence of leakage of contrast material at the grafted/implanted site in the five groups, whereas in two animals of group E, where recombined ceramic implant with marrow is used, leakage of contrast material shows from the implanted site.

Lead oxide-soap suspension is (20%) provides excellent contrast and vascularisation of the arterial system even up to the capillary level. The course and distribution of the vascular system can be compared with normal angiography in all six groups at day 90 postoperatively. Vessels are patent, well defined, organized and uniformly filled with contrast material. The fine branches of periosteal and endosteal vessels penetrating the bone graft are discernible in groups A, B and C. Endosteal vessels invading the graft bed are greater in number than periosteal vessels; there are no signs of capillary regression in any group. Angiograms taken after removal of soft tissue show that cross-over of medullary vessels across the grafted site are more prominent in groups A, B and C than in groups D, E and F. In general, the intensity of vascularisation is greater in groups A, B and C than in groups D, E and F.

Photographic enlarged undecalcified sections reveal that the bone defects are almost completely bridged by newly formed osseous tissue in groups A, B and C. In group A, the bridging is complete on endosteal side whereas discontinuity of bone is seen on periosteal side. In group B, bone defects are completely bridged by new bone originating from both the endosteal and periosteal sides, but in group C, the amount of new bone formation is comparatively less than that of group B. The photographic enlarged undecalcified sections show that the bridging of the bone defects is partial /incomplete in the groups of D, E and F, where ALCAP ceramics alone and in combination are used respectively. In all these groups, the intensity of new bone formation is mild to moderate. In some sections, the defect is bridges on its endosteal side and lipping of newly formed bone is seen on periosteal side. Most of the implants are lost during grinding process suggesting their loose attachment at the implanted side. The bone surface lining the empty space is uniform and smooth suggesting failure of new bone ingrowths into the implant.

In animals of group A, the triple fluorochrome-labeled ground sections reveal that the newly formed bones originated from periosteum and endosteum surface with higher intensity from periosteal side. The bone defect is completely filled with newly formed cancellous bone. At places, remnants of graft are still present with signs of disintegration and fragmentation. Cavities formed due to resorption are lined by newly formed osteoid tissues. At places, indications of osteogenic activity in the decalcified graft are existed. In group B, the amount of new bone formation is more as compare to group A. The bone defect is almost completely filled with newly formed bone and the graft is in the process of resorption. Graft is appeared as fragmented homogenous mass between the trabeculae of newly formed osseous tissue. The presence of bright yellow organized areas within the graft is suggestive of osteogenic activity in the graft. New bone formation is moderate in the plasma-coated decalcified allograft of group C. The newly formed bones penetrate the graft from all the sides resulting into bridging of the bone defect. The activity of new bone formation is mild to moderate adjacent to the implant sites in groups D, E and F. In most of the cases, implants are lost leaving a gap apposed by newly formed osseous tissues. The gap is mainly bridged on endosteal side. On periosteal side lipping of new bone is seen on both the side which failed to unite. In these groups, new bone originated mainly from the host bone and failed to penetrate ceramic implants, showing poor osteoinduction and osteoconduction of ALCAP ceramics.

Histologically, at day 90, the process of new bone formation is distinctly noticed in frozen decalcified allograft group. The union of the graft with host bone is observed conspicuously at their borders, which is noticed as a wider bluish tinged zone. Resorption and replacement process started within the peripheral clefts of the decalcified bone grafts. Presence of osteoclasts, widening of the Haversian canals at the border of the graft of group A, suggests the process of resorption, followed by new bone formation as evidenced by the attachment of graft with host trabecular bone on both endosteal and periosteal sides and presence of blood vessels entering into the resorption cavities formed within the graft. In some of the areas, the presence of the Haversian system cannot be visualized properly as a result of the abundant deposition of the new bone in the inter-Haversian spaces as well as widened Haversian canals. The resorption of recombined decalcified allograft with autogenous red bone marrow (Gr B) is more rapid than that of decalcified allograft alone (Gr A). The presence of varying sized resorption cavities containing trabecular bone with newly formed blood vessels and evidence of extensive osteoclastic activity is the noticeable feature in these grafts. At places the graft is having an appearance of fibrocellular network with the presence of islands of old grafted bone, indicating the extensive resorption and replacement process in the grafts. The resorption and replacement of the plasma coated allograft (Gr C) is appeared to be comparatively slower that of Gr B. The presence of the graft in its original place can be viewed and identified by the presence of poorly organized lamellar bone with ill defined Haversian system containing empty osteocytic lacunae and areas of osteoid tissue. The histological examination of ceramic implants is not possible in this study as the unresorbed implants are detached during the processing on account of their weak union with the host bone. However, mild to moderate new bone formation from the adjacent host bone is evident in all the three groups where ALCAP ceramics alone (Gr D) or in combination with autogenous bone marrow (Gr E) and autogenous plasma (Gr F) are used.

In conclusion, the radiological, osteomedulography, angiography, enlarged undecalcified ground section, triple fluorochrome and histopathological observations are confirmed that comparatively faster new bone formation in the composite bone grafts and presence of bone matrix is must for utilization of osteogenic property of bone marrow and citrated plasma. ALCAP ceramics are not effective as decalcified bone graft. Osteomedulography, angiography, undecalcified ground section studies are important parameters for bone healing study. Triple fluorochrome labeling study is also find a very useful to detect the amount, source of new bone formation, healing process and extent of new bone formation in the evaluation of different types of bone grafts and ceramic implants.

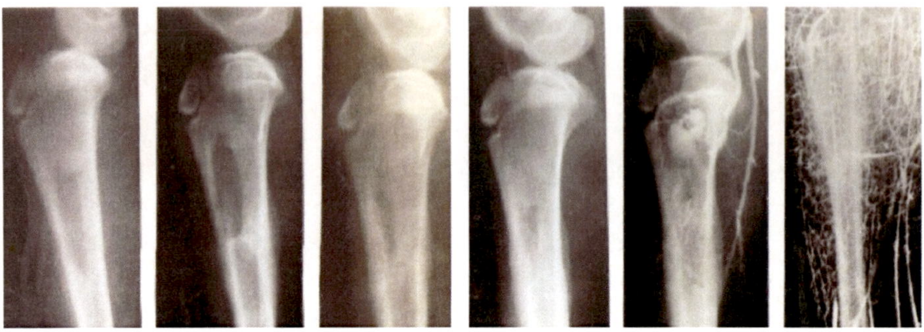

Radiography, osteomedulography and angiography of Group A

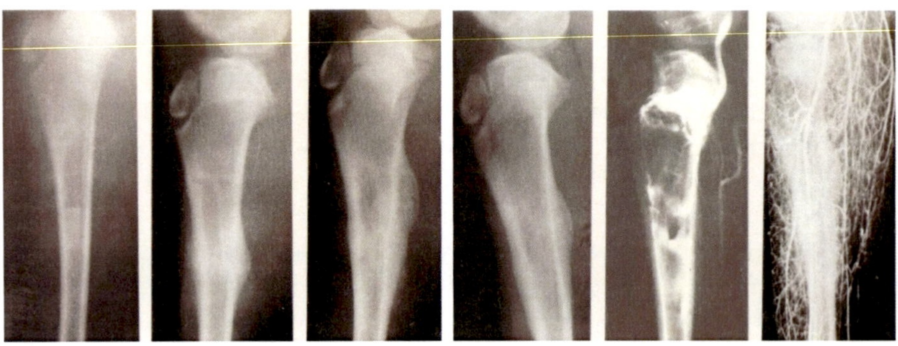

Radiography, osteomedulography and angiography of Group B

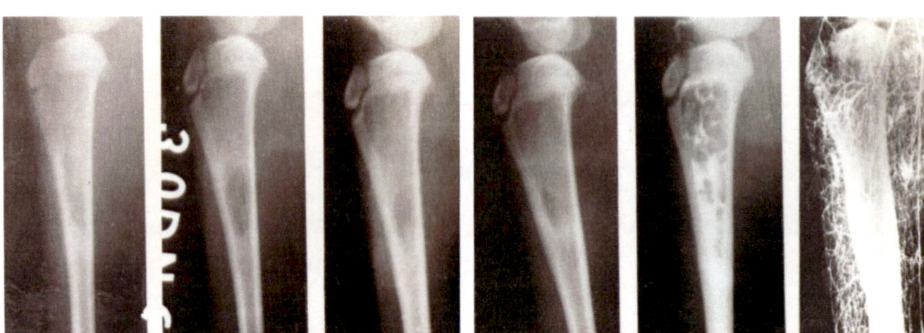

Radiography, osteomedulography and angiography of Group C

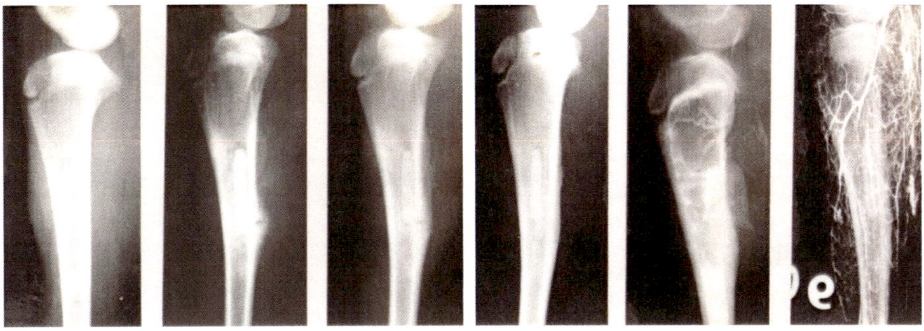

Radiography, osteomedulography and angiography of Group D

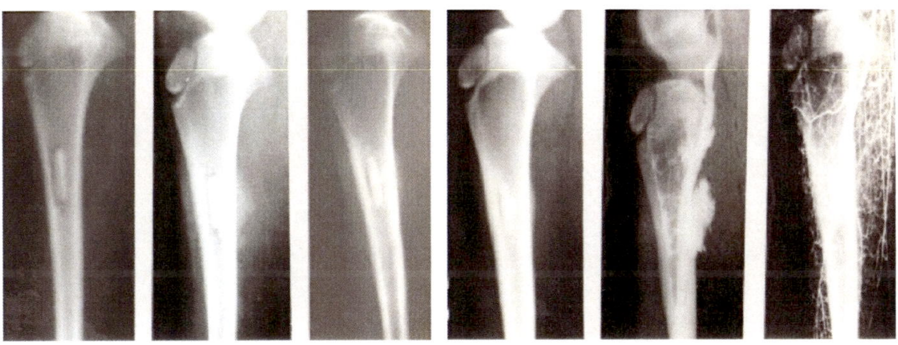

Radiography, osteomedulography and angiography of Group E

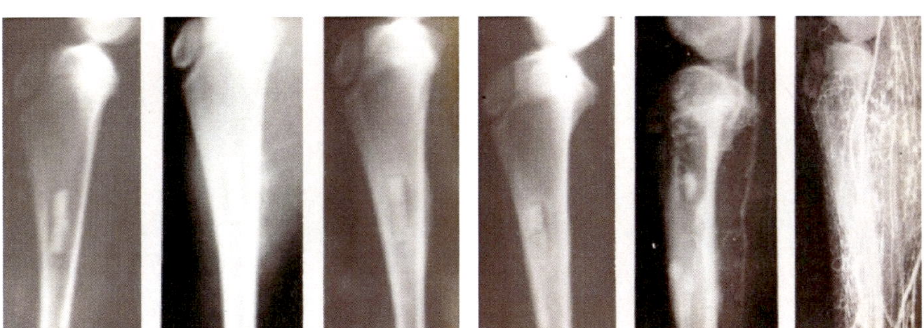

Radiography, osteomedulography and angiography of Group F

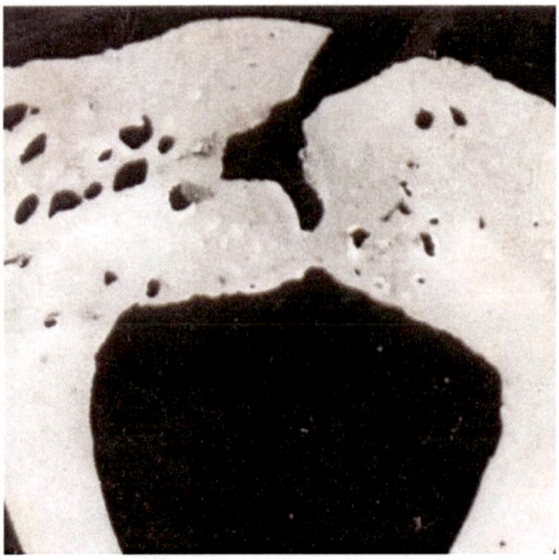
Undecalcified ground section of Group A

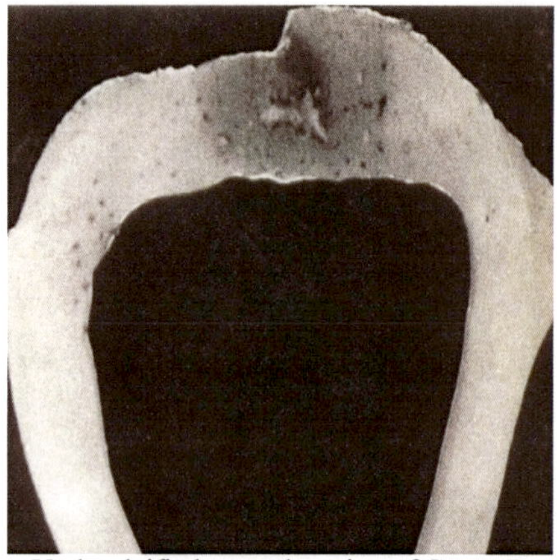
Undecalcified ground section of Group B

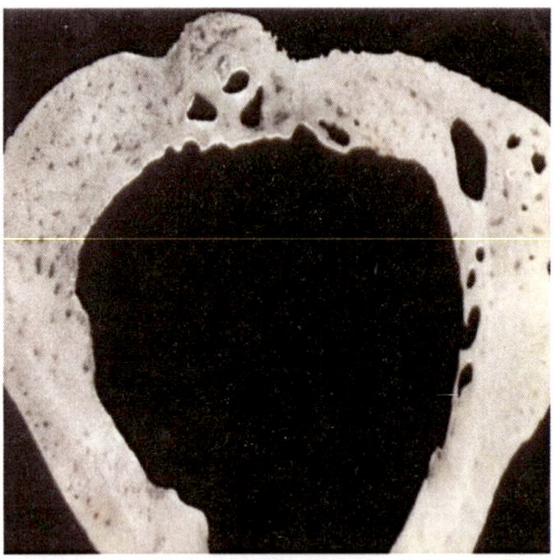
Undecalcified ground section of Group C

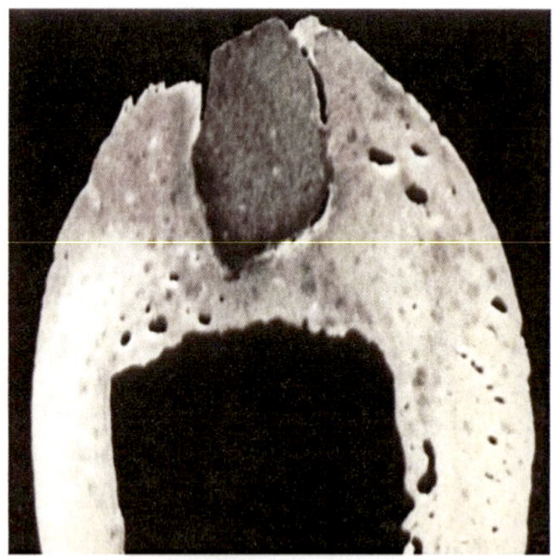
Undecalcified ground section of Group D

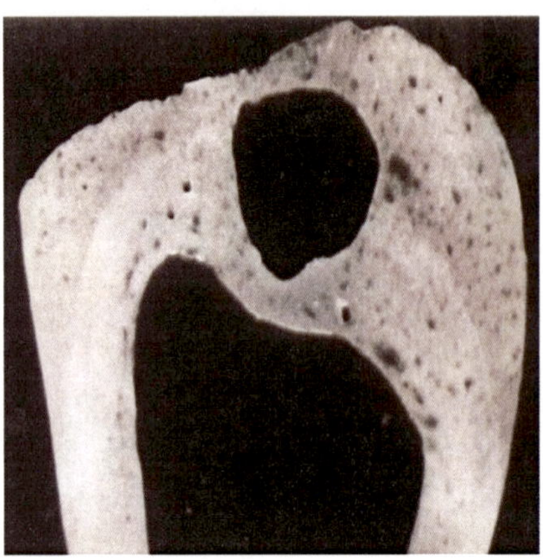
Undecalcified ground section of Group E

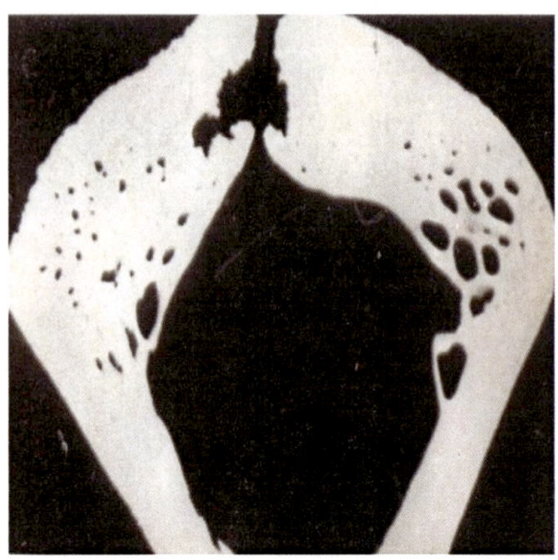
Undecalcified ground section of Group F

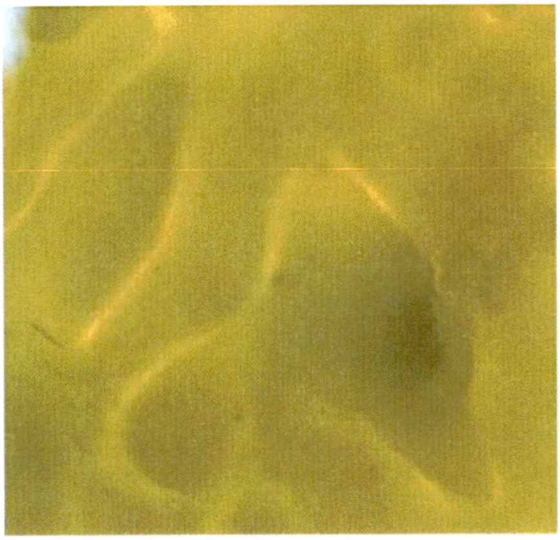

Triple Fluorochrome labeled photomicrograph of Group A

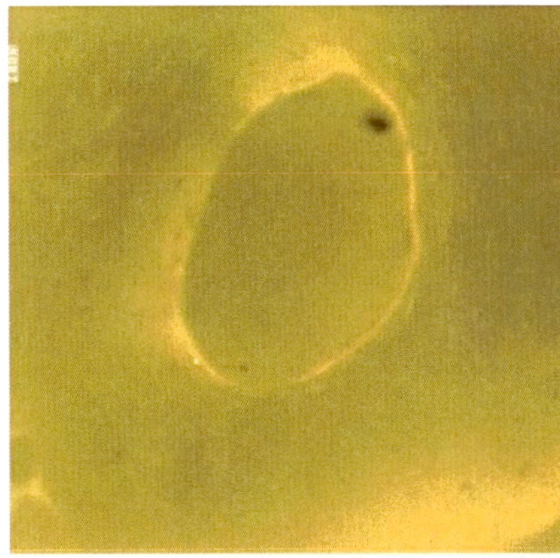

Triple Fluorochrome labeled photomicrograph of Group B

Triple Fluorochrome labeled photomicrograph of Group C

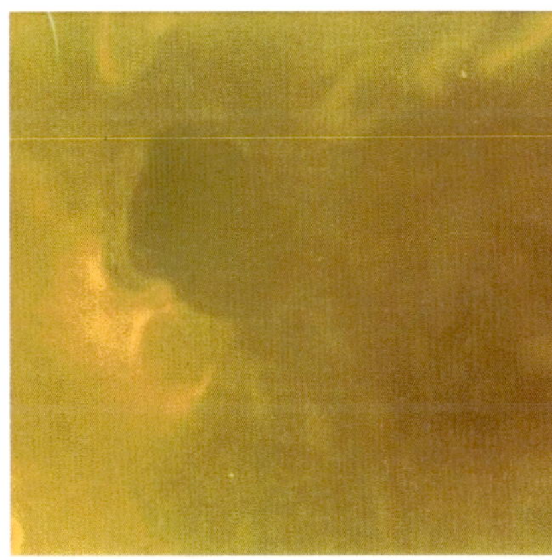

Triple Fluorochrome labeled photomicrograph of Group D

Triple Fluorochrome labeled photomicrograph of Group E

Triple Fluorochrome labeled photomicrograph of Group F

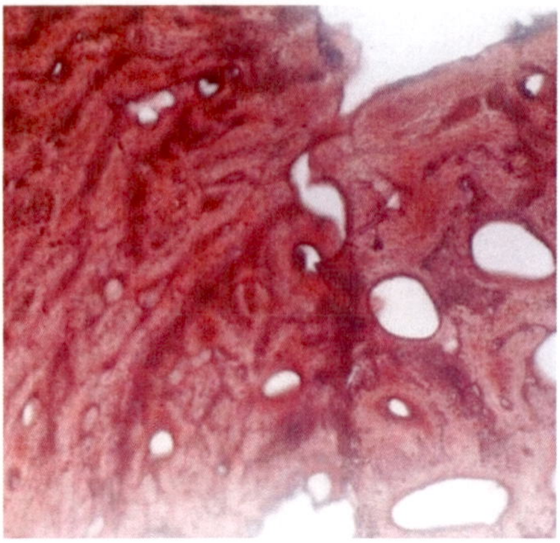

Histopathological photomicrograph of Group A

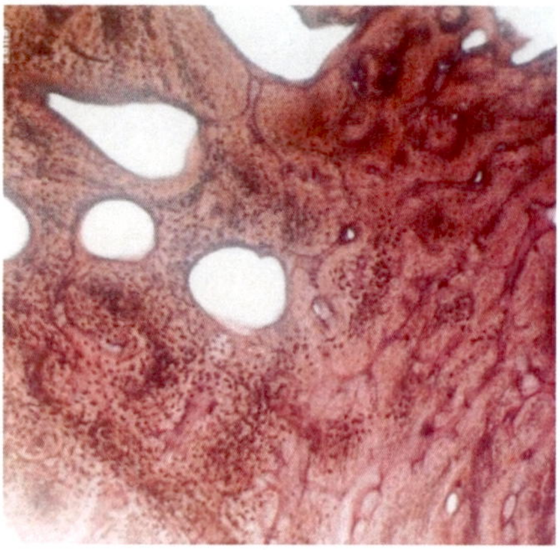

Histopathological photomicrograph of Group B

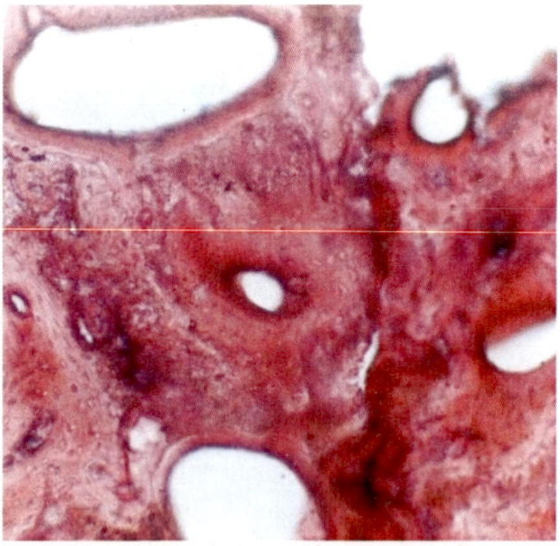

Histopathological photomicrograph of Group C

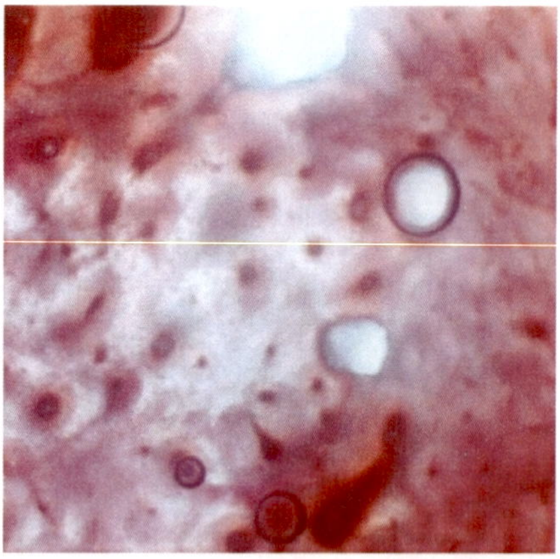

Histopathological photomicrograph of Group D

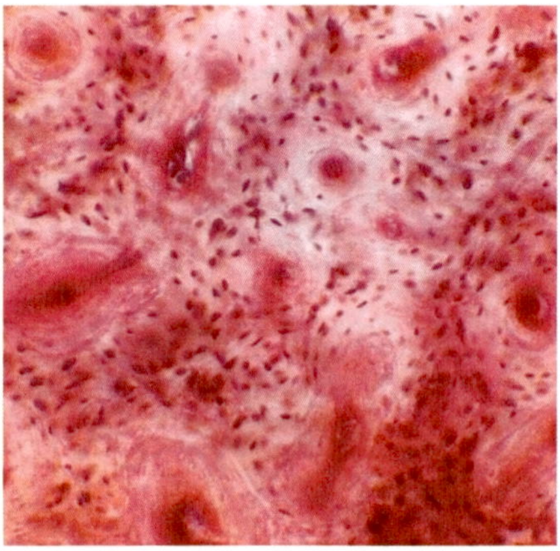

Histopathological photomicrograph of Group E

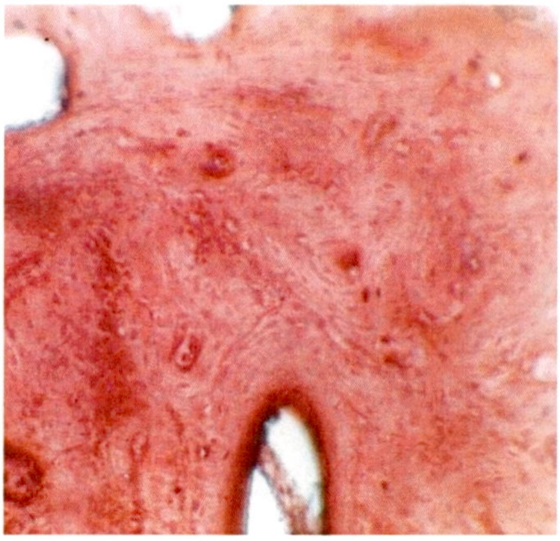

Histopathological photomicrograph of Group F

27

Physiotherapy and Postoperative Rehabilitation for Orthopedic Patients

Physiotherapy may be defined as the use of physical techniques like massage, exercise, heat, cold, sound waves, electric current etc for the treatment of injuries and movement dysfunction. Physiotherapy deals with the treatment of diseases by physical methods. It is used for treating injuries, strain, spasm, painful conditions, sprain of tendons, paralysis, postoperative trauma etc. It improves functioning of nerve, muscles and bones. The aim of physical therapy is the restoration of function and promotion of tissue healing by assisting normal physiological processes. The physiological response to physical therapy is its effect on the vascular supply; this in turn reproduces similar changes in deeper tissues.

Common benefits of physiotherapy for animals

- Improves function
- Improves endurance
- Improves biomechanics of joints
- Improves posture
- Improves balance
- Improves proprioception
- Increases strength
- Increases range and quality of motion
- Minimizes muscle atrophy following surgery or injury
- Reduces pain and discomfort
- Reduces muscle tightness and spasms
- Increases flexibility, thereby reducing the risk of injury
- Improves and speeds recovery from surgery or injury

- Minimizes or eliminates complications following surgery or injury
- Improves neurological conditions - speeds healing and return to function/ walking.

Indications for physiotherapy in orthopedic conditions

- Arthritis
- Hip dysplasia, elbow dysplasia
- Patella luxations
- Fragmented choronoid processes
- Limitations in range of motion, contractures
- Post-fractures
- Optimize repair of fractures both normal and non-union

Types of Physiotherapy

Physiotherapy encompasses many different techniques. These include both the skilled use of the physiotherapist's hands and the use of modern therapeutic modalities:

- Therapeutic exercise
- Massages and Passive movements
- Heat and Cold
- Electrical stimulation
- Diathermy
- Therapeutic Ultrasound
- Phonophoresis and Iontophoresis
- Transcutaneous Electrical Nerve Stimulators (TENS)
- Interferential therapy
- Hydrotherapy / Swimming/Aquatic therapy

A treatment may involve just one or a combination of the above techniques.

The goal of each treatment is vary according to each individual case, but generally is include:

- Reduction of pain
- Improvement of mobility
- Return to pre-injury strength and flexibility
- Prevention of reoccurrence of injury

Therapeutic Exercise

Of all the modalities used in small animal's physical therapy, therapeutic exercise is often the most effective in cost and achievement. Therapeutic exercise can be used to preserve passive range of motion and muscle mass and to challenge healing tissues during recovery. Exercise is frequently used in subacute and chronic conditions and it is used to remove swelling, particularly in puncture wounds of the limb, midline incisions, castrations and post fracture immobilization. It should be combined with massage and the uses of liniments. It may also be used to rehabilitate and strengthen the limb after injury to tendon or ligament. The goal of therapeutic exercise is a return to full functionality and normal range of motion. Because each animal is an individual, no "Cook-book" recommendation can be made. However, the exercise program may include some of the following exercise:

- Weight shifting
- Step up and step down
- Step over
- Serpentines
- Circles
- Figure 8's
- Transitions (Slow-fast-slow walk; walk-trot-walk)
- Leash walking on level and uneven surface
- Swimming

An important factor to acknowledge is the need to match the specific injury/ and or surgical repair with the appropriate exercise. Following cranial cruciate ligament repair in dogs, the quadriceps, biceps and semi-membranous muscle undergo significant atrophy. Therapeutic exercise plan for a postoperative cruciate patient, concentration should be given specifically on these muscles.

A controlled exercise programme is based on the principle of progressive resistance. The exercise regimen begins with the least stressful and the most controllable form of activity. Rehabilitative exercise should be tailored to individual needs in terms of duration and intensity to be of benefit. The use of a heart rate monitor or thermograph gives a better indication of exercise stress or benefit. Physioballs, treadmills, leg weights or swimming are more effective than hand walking and automatic walkers because they provide the animal with a means to exercise without the concussion of work on the ground. The advantages of these tools are that weight bearing is decreased and the resistance of the water stimulates the cardio-respiratory and muscular systems.

Physioballs are available in various sizes. Dogs can be positioned with their thoracic limbs resting on the Physioballs in order to increase weight bearing on the pelvic limbs. Care should be taken to avoid over-extension of the spine, which potentially could stress the lumbosacral spine, coxofemoral joints and/or sacro-iliac joints.

Treadmills are available in various sizes for dogs. The moving belt of the treadmill ensures that the animal must move using all four limbs evenly in order to maintain in balance. It is very good for rehabilitation of gait pattern for dogs and horses. Treadmills are convenient and allow the trainer to control the exercise with respect to time, speed and gradient. The treadmills should be cover with a wall facing the moving animal. This prevents distraction and reduces the risk of the falling or slipping. The disadvantage is that treadmills do not simulate normal movement - the ground does not usually move under the dog's feet! Consequently, the pattern of muscle activation may differ when the dog is exercising on a treadmill. Therefore, it is recommended that an exercise program incorporate other exercise in addition to treadmill work.

Swimming-It is a non-weight bearing exercise in which the limbs work against the resistance of the water in a symmetrical pattern. Swimming is an excellent way to rehabilitate the injured animal without concussion to the limbs. It should be used whenever possible to strengthen the limbs after surgery before regular training procedures are reinstituted. The animal should swim in straight lines. When animal shows the signs of exhaustion, swimming should be stopped. Swimming for the first circuit should be short followed by a period of rest and then start for second circuit of swimming. Heart and respiration should return normal before the start of second circuit.

Leg weights-Injury to muscle or ligament may cause muscle wastage and muscle atrophy. To re-build and gain the pre-injury state of the muscle and ligament needs an increased workload by applying weights in the injured limbs. For this , a 5-7 oz (as per dog size) strip of lead may bind with the masking tape and fix it around the injured leg. This leg weight should be used for 10-15 minutes daily for 10-25 days as indicated by the improvement of the lesion.

Massage

Massage is 'moving" the soft tissues in the body with hand or machines. Manipulation of soft tissues with hands or fingers is called massage. Massage improves circulatory flow, increase nutritional supply, removes waste products, reduces edema and indurations and muscle spasm, mobilizes the adherent tissue and loosen the tight and contracted tendons, ligaments and muscles. Massage, in general, should be done in the direction of the venous flow. When massaging an extremity, it may be elevated when possible to make use of gravity in assisting venous return. For massage application, the skin over the target tissue should be intact. The motion should be rhythmic and pressure is moderate. Massage should be administered skillfully. The therapist should have much patient, perceptive and sensitive with the hands. The therapist should start massage away from the injured area to avoid creating guarding muscle contractions. Frequent and short massages are better than long, irregular and infrequent sessions.

Indications

- Soft tissue swelling
- Muscle tear/injury/soreness
- Ligament/muscle sprains/pull
- Muscle tension/spasm
- Osteoarthritis
- Bursitis
- Tendinitis

Contradications

- Infection/sepsis in an area
- Fracture site
- Metastasis disease
- Acute traumatic conditions
- Torn muscle
- Hemorrhage
- Presence of foreign bodies
- Presence of phlebitis
- Presence of acute hematomas
- Presence of inflammatory arthritis
- Recent high fever
- High blood pressure

Different kinds of massages commonly employed are:

Hand massage: The palms of the hands are mould according to the contour of the affected area. For this, use of relaxed hands is advocated, the hands should direct to influence venous return, and linear/circular/direct compressive strokes should be applied. It is done 10-30 times daily. Olive oil, mustard oil, coconut oil or a fine-grained talcum powder can be used to reduce friction, to soften skin and avoid pulling of hair.

Friction/stroke massage: Small and deep strokes by fingertip across and linear to tissue fibers are applied should be followed by general massage. The strokes should be of even pressure and run longitudinally along the muscle, following the direction of hair growth. The initial stroking should be superficial rather than deep to create a sedating effect and prevent protective muscle contractions. The strokes of a massage given for sedative purpose should be slow and rhythmic. Special strokes are used to stimulate. Random movements of the hands are ineffective. Massage requires concentration on each manipulation. It should be done for 3-7 minute followed by 15-20 minute massages on alternate days. Stroking is useful to begin and end a massage session.

Ice massage: Ice massage involves massaging a block of ice directly to the skin or cube frozen around a 'lolly stick'. A convenient method of applying ice massage is to freeze water in a Styrofoam cup and then hold the cup base and apply the ice massage directly to the injured area for 5-10 minutes depending upon the size and intensity of the inflammation. Cold lowers cellular metabolism and causes vasoconstriction and lowers nerve conduction velocity, which ultimately leads to increase pain threshold value. Cold penetrates deeply to heal deeper tissues lying under skin. Ice massage is very useful in any acute swelling, muscle soreness, tendinitis, bursitis, and arthritis and ligament injury.

Machine massage: It effects vibrations rather than massage, requires no experienced hands. For this, the animal should be accustomed to noise of the machine. The strokes should be linear to fibers with firm and even pressure. It can be done 20-30min twice daily. For application of the machine massage, animal should be controlled or restrained either in standing or recumbent position as per requirement. Animal should be accustomed to both the noise and vibration of the machine. Machine should be moved in the linear direction of the fibers of the underlying tissue, maintaining a standard pressure. Higher pressure should be applied over areas of muscle bulk than over bony prominences. Massages should be done for 15-30 min, twice a day or as per requirement. The effect is transitory, and treatment must be repeated several times a day to be of most value.

Massage therapy is one of the ways to free up tight muscle and enhance blood circulation.

It also has many more benefits such as:

- Relaxes muscle spasm
- Improves muscle tone

- Reduces congestion within blood and lymph vessels
- Increases excretion of fluids
- Increases the exchange of substances between cells
- Encourages tissue repair
- Stretches connective tissues
- Helps maintain flexibility
- Reduces inflammation of tissues in joints
- Increases intestinal movements
- Stimulates glands
- Improvement of blood circulation
- Reduction of swelling/edema
- Transient loss of pain
- Mobilization of adherent tissue
- Stretching of contractions
- Freeing scar tissue adhesion of the skin to underlying tissues

Liniments & Rubs: Massage is used in sub acute and chronic swelling and can be combined with the use of liniments, although the main benefit is from massage. The lubrication quality of the liniment aids in the massaging. Many liniments receive credit for being good therapeutic agents, when it is actually the massage that reduces edematous swelling and aids in relieving pain of an injured tendon or joint. Liniments should be applied before massage. It increases blood flow. The effect is temporary and it should repeat several times a day for getting better result. There are many types of liniments are available in the market. Among them, olive oil, coconut oil, mustard oil, aloe Vera, arnica, turpentine oil, Arachis oil, corticosteroids, salicylic acid, heparin, lignocaine/xylocaine jelly etc are used for this purpose.

Passive Movements

It is a good means of regaining joint mobility. It shows good results when being done after heat and massage. Passive motion involves repeated flexion and extension of affected joint without weight bearing. It should be done in full range of mobility of joints with gentle overpressure, with due care not to cause any injuries to the joints. No muscle action is involved in this exercise. These movements of a specific joint are done without voluntary assistance or resistance from the animal. To perform passive movements correctly, the therapist must be aware of the normal range of movement of the particular joint to be treated.

Passive movements are performed within the limits of the soft tissue without stretching. The benefits of passive motion are the improved health and healing of the intra-articular structures particularly the cartilage and reduction of soft tissue contracture associated with immobilization. Metabolic activity in the cartilage is improved and postoperative fibrosis of periarticular structures is reduced when repeated gentle movements are manually assisted in the affected areas. Passive movements improve synovial fluid production in the joints and increases joint mobility. Passive motion should not be performed until the surgical incision is healed sufficiently.

Indications

- Joint inflammation
- Osteoarthritis joint disease
- Joints with restricted movement following trauma

Contradication

- Congenital dislocations of joints
- Hip dysplasia
- Articular fracture

General Effects

- Prevent adhesion
- Maximize the range of joint motion
- Improvement of circulation
- Enhance rate of healing
- Rapid resorption of waste products
- Increases bone response to mechanical stimulation
- Reduce bone atrophy caused by decreased limb use

Procedure

- The proximal end of the limb is held firmly with left hand above the affected joint.
- The distal end of the limb below the affected joint is held with right hand and gently flexion-extension (full range) movements are given
- If restriction of movement is present, mild "over pressure" is applied at the end of the movement.

Continuous Passive Movements (CPM) will stimulate joint proprioceptors and helps in the regeneration of the primitive articular cartilage.

Thermal Agents

Superficial heating agents penetrate to a depth of approximately one centimeter. Deep heating agents can elevate tissue temperatures at depths of three centimeters or more. Deep heating agents include ultrasound and diathermy. As a general rule, cold therapy ("cryotherapy") is indicated during acute inflammation when the lesion exhibits signs of redness, swelling, pain and/or heat. Therapeutic heat is indicated during the sub acute and chronic inflammatory phases. Ice application is also indicated to resolve swelling after exercise. The indications for alternating heat and cold ("contrast baths") include impaired venous circulation and indolent ulcers, and traumatic or inflammatory conditions during the sub acute or chronic phases. The mechanism of action of contrast baths is production of alternating vasoconstriction and vasodilatation of local blood vessels. This reaction is considered to stimulate blood flow to the treated area and thereby to help stimulate healing.

Therapeutic Heat (Thermotherapy)

Heat can be applied as radiant heat, conductive heat and conversive heat. Physiological effects of the three methods are basically the same. Radiant heat is applied by infrared light, conductive heat by hot water bottle, electric heating pad, hot fomentations and poultices; conversive heat is by diathermy and therapeutic ultrasound. Conversive heat is generated in tissues by resistance to high frequency electrical energy-diathermy or sound waves-ultrasound. Heat is generally used to increase circulatory response for resorption of swelling developed by accumulation of blood or serum. With heating, smooth muscle tone decreases while vessel diameter increases, resulting in greater perfusion of the tissue. It causes vasodilatation due to release of prostaglandin and histamine in the applied area. Heat-induced local vasodilatations increase the local blood flow 4 to5 times that of the resting state. Heat increases phagocytic activity, oxygen supply, lymph flow, and cellular metabolism in the applied area. As heat causes vasodilatation, it also increases non-significantly the local temperature. Heat also increases capillary permeability which causes increased absorption of endogenous toxins developed by tissue injury.

Indications

- General injury
- Following surgical intervention
- Post trauma
- Muscle/ligament sprain/stiffness

Contradications

- Infection
- Recent injury (24-48 hrs)
- Metastasis disease

Therapeutic heating devices are classified into two groups: superficial and deep. Superficial heating devices includes hot packs, heated pads, hot water bottles, turbulator, Infra red lamps and sunlight. They primarily increase cutaneous and subcutaneous temperature and have little effect on the deeper tissue. Deep-heating devices include diathermy and therapeutic ultrasound. These deep-heating devices have the ability to heat tissues 3 to 5 cm below the skin. For heating devices to be effective, temperature must increase in the target tissues, not just on the skin surface. In an effort to heat deep tissues, therapist generally overheating the superficial tissues. Increasing the temperature of these devices does not increase their depth of penetration; it only increases the danger of thermal damage to the skin. Care should be necessary during heat therapy particularly for deep heat.

Therapeutic Cold

The application of cold is used in treatment of acute and hyper acute inflammatory processes. It helps in relief of pain and it is used to prevent edema and tissue swelling. There is a reduction in tissue metabolism rate and it also causes surface anesthesia. The best result is obtained when cold therapy combined with compression bandage and rest to further limit the swelling. Cold therapy should be applied within 24—48 hrs after trauma. After this time, it is of no value. Duration of cold therapy is generally 30-40 min, then one-hour interval and then reapplied.

Cold causes immediate vasoconstriction in the peripheral parts of the body thereby causing decreased exudation. This is followed by vasodilatation of deep arteries as part of the thermoregulatory mechanisms of the body known as 'Hunting response'. Cold reduces the oxygen requirements of the uninjured tissue thereby making them competent.

Methods of Cold Therapy

- **Running stream-** stream of running cold water applied to the affected area or limb. This type of water therapy should not be used in case of presence of open wound.
- **Whirl pool tubs-** tubs with water current generated by electricity and filled with crush ice up to knee/hock

- **Welli boots**- special boots containing crush ice inside. It is expensive.
- **Hosing-** the most common method is the application of cold water with a hose. The water is usually colder than the body part being treated, but the water temperature and the accuracy of application are uncontrolled. The massaging effect of the water stream can further damage exposed sub-dermal cells and capillary. It is usually applied for 30-45 minutes.
- **Cold bandages**- applied to treat limbs. It can remain at 6°C after 10 minute freezing. These are commercially available and applied as per manufactures instructions.
- **Cotton wool pads**- mould is prepared with dampens cotton, freeze in the shape of affected part and apply.
- **Ice massage**- this technique can be used for treatment of any parts of body. The application of ice and a compression wrap is a better method of cold therapy. The cold wrap should be removed after 15-20 minute and a compression wrap should be applied again for extending the period of ischemia. An ice cubes also can apply for a local massage. A convenient form of cold application is the ice-cup-massage. A Styrofoam cup filled with water and then frozen provides an effective ice-therapy tool. Ice massage is given for 15-20 minutes many times a day and combined with compression wraps. Ice massage is more effective than the use of crushed ice packs because the air surrounding the pieces of ice results in uneven applications.

Indications

- Acute trauma/ inflammation
- Hyper acute inflammation
- Local oedema
- Local haematoma
- Injury of muscle or ligament or joint.
- Laminitis

Contradications

- Open injury
- Fractures
- Infection/sepsis
- Large wound

General Effects

- Immediate vasoconstriction followed by deep tissue vasodilation
- Reduce oedema/swelling
- Reduce tissue temperature
- Reducing irritability of the nerve-muscle junction
- Reduction of muscle spasm
- Early movement possible
- Reduce tissue metabolism
- Decrease rate of local tissue damage
- Increase deep circulation
- Local anesthesia
- Enhance rate of tissue healing

Contrast baths: The indications for alternating heat and cold ("contrast baths") include impaired venous circulation, indolent ulcers, soft tissue damage, haematoma and traumatic or inflammatory conditions during the subacute or chronic phases. The mechanism of action of contrast baths is production of alternating vasoconstriction (due to cold) and vasodilation (due to heat) of local blood vessels. This reaction is considered to stimulate blood flow to the treated area and thereby to help stimulate healing. Contraindications of contrast bath are open wounds, infection and sepsis.

Diathermy

Diathermy will penetrate to a depth of approximately 2-inch. There is a problem of arcing and electrical shorting in treating animals.

(I) Short wave diathermy

Short-wave diathermy units are available in the market in which the current oscillates several million cycles per second. Short-wave diathermy produced heat in the tissue by means of high frequency alternating current, which is converted to heat without inducing muscular contractions. The tissues heats because of resistance to this energy and it are unable to expel the heat produced. It utilizes high-frequency alternating currents that oscillate between 10 to 50 MHz the value of this modality is that these effects of any heat therapy, except these effects occur in the deep tissues. Heating by high-frequency electrical energy is obtained by including the tissues in the circuit. The high-frequency electrical energy passes from one electrode through the tissues to the other electrode. Resistance in the tissues produces a variable rise in tissue temperature due to a variation in the degree of impedance to the electrical energy. Therefore, all tissues will not be heated to the same degree. The greater the fluid content of the tissue, the higher the temperature recorded. As fats are more resistance, they damaged earlier than other tissues. Bone and tendon have lower water content; they do not heat as much as the surrounding moist tissues. This therapy should not be given if metal implants are present. If bone screws or intramedullary pins or bone plates are present in the bone, they may heat to a high temperature leads to bone necrosis. Treatment time is generally given 15-20 minutes on alternate days. However, diathermy treatment can be used twice daily if necessary. The therapeutic effect of a lower dose given over a longer period of time is better than that of a more intense dose given over a shorter period.

(II) Microwave diathermy

The general effects are similar to that of short-wave diathermy. It involves the use of high frequency electromagnetic radiation that is absorbed by body tissue and produces molecular agitation and thereof thermal energy. There is no radiation damage. The microwave penetrates up to 3-5 cm in the soft tissues. The treatment time is 15-20 minutes. Microwave therapy is not always superior to other forms of heat therapy except for its immediate deep effect.

Indications

- Soft tissue injury
- Joint injury
- Tendonitis
- Myositis
- Adhesion
- Para vertebral pain
- Muscle spasm
- Joint stiffness following immobilization
- Articular/periarticular scars

Contradications

- Presence of metal implants
- Periocular region
- Infections
- Malignancies
- Idiopathic hemorrhagic problem

- Pregnancy
- Nerve injury
- Spinal affections

Therapeutic Ultrasound (TU)

It is over 70 years since the interaction between high (ultra) frequency sound waves and living tissue are initially studied and Wood and Loomis (1927) first suggested the use of such energy as a form of therapy. Therapeutic ultrasound becomes now one of the most commonly used treatments in the management of soft tissue injuries. Ultrasound is a form of mechanical energy, not electrical energy. Mechanical vibrations at increasing frequencies are known as sound energy. Below about 16 Hz, these vibrations are not recognizable as sound, and the normal human sound range is between 15-20,000Hz. Beyond this upper limit, the mechanical vibration is known as ultrasound. The frequencies used in therapy are typically between 1.0 and 3.0 MHz (1MHz = 1 million cycles per second).

The thermal effects of US are:

1. It increases metabolic rate of tissues.
2. It increases blood and lymphatic flow.
3. It increases extensibility of collagenous tissues (tendons, scars, muscle sheaths, joint capsules, and ligaments).
4. It alters the elastic properties of collagen tissue and its molecular bonding.
5. It changes in nerve conduction velocity.
6. It decreases pain and muscle spasm.

The non-thermal effects of US are:

1. It increases cell diffusion and cell membrane permeability.
2. It increases production of collagen and hydroxyproline.
3. It increases fibroblast proliferation.
4. It stimulates tissue repair.
5. It increases tensile strength and energy absorbing capacity of tendons and ligaments

Therapeutic ultrasound is effective, non-invasive, and affordable. The treatment can be administered directly to the affected area, penetrating deeply into the tissues. The rise in temperature increases blood flow, relaxes muscle spasms, helps repair damaged tissue, and speeds healing dramatically. It is the only modality known to effect permanent, physical changes and to repair both hard and soft tissue problems.

Indications

- Tendon and ligament Injuries
- Soft tissue and muscle Injuries
- Scar tissue and contractors
- Back conditions, spondylosis
- Hind quarter weakness
- Joint problems, arthritis, bursitis, post surgery

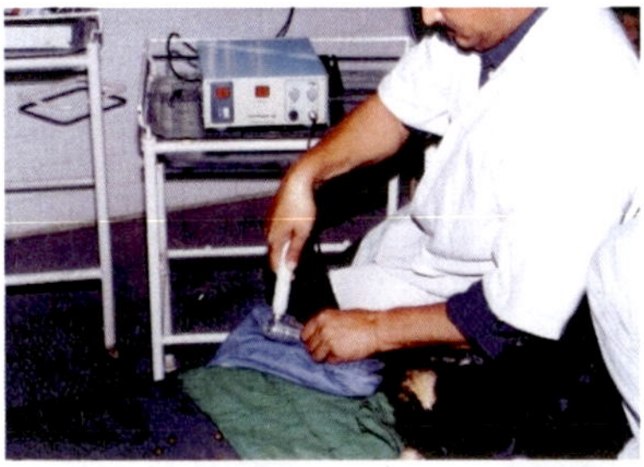

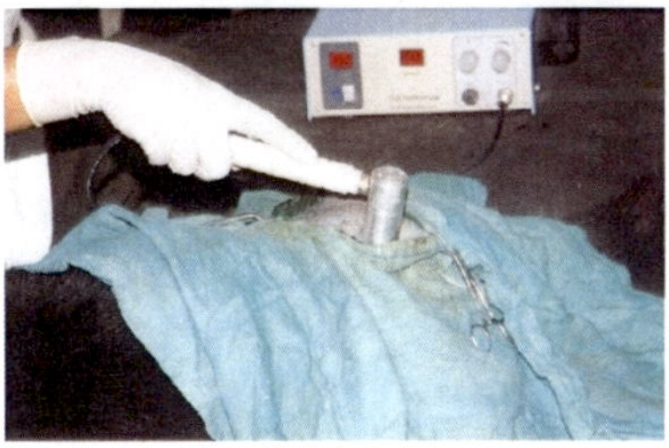

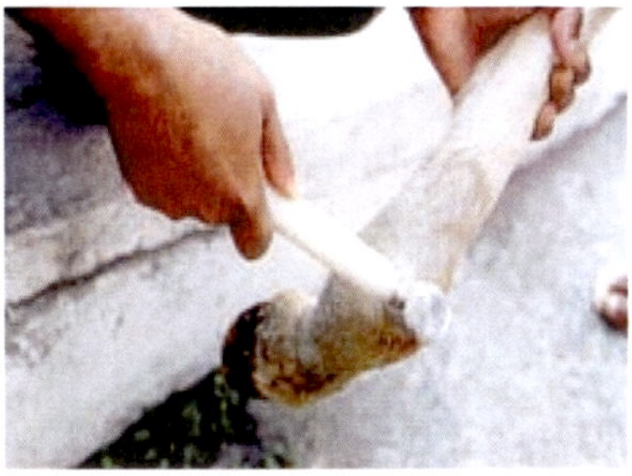

Therapeutic Effects

- Reduces calcium deposits
- Breaks up and softens scar tissue and adhesions
- Stimulates healing without irritation
- Speeds metabolism and improves blood flow
- Reduces nerve root irritation
- Enhances the body's natural healing process
- Ultrasound energy absorbed at molecular level
- Proteins of the soft tissues are major absorbers
- Nerve protein is sensitive to ultrasound
- Blood haemoglobin absorbs ultrasound
- Cell membrane absorbs ultrasound
- Muscle protein absorbs ultrasound

Dosage

Ultrasound dosage represents the rate at which it is being delivered per unit area of the transducer surface is Watt/cm^2. It can change with alteration of wave amplitude and intensity. For ultrasound dose calculation, the first steps involve the decisions as to which machine settings are most appropriately applied to the patient's particular problem. The second stage is to bring these into an effective treatment combination. **Machine Frequency:** Taking into account that the most frequently available ultrasound machine, ultrasound frequency is 1 and 3 MHz, the option between them relates primarily to the effective treatment depth that is required. **Pulse Ratio:** The pulse ratio determines the concentration of the energy on a time basis. The pulse ratio determines the proportion of time that the machine is ON compared with the OFF time. **Ultrasound Intensity**: The intensity of ultrasound (Watt/cm^2) required on the target tissue will vary with the tissue state. The more acute the lesion, the smaller ultrasound frequency, whereas, more chronic the lesion, greater the intensity is needed. **Size of the lesion:** The greater the size of the lesion, the longer the duration of the ultrasound will be required. The most common method to take account of this factor is to estimate the number of times which the ultrasound head to be utilized can be placed over the target tissue.

Therapeutic ultrasound can be pulsed or continuous. Continuous ultrasound has a greater heating effect whereas pulsed ultrasound produces non-thermal effects. There are no set dosage criteria for therapeutic ultrasound. The condition must be analysis regularly and personal experience is the best guide. Duration of treatment depends on the condition treated and the size of the treatment site. Treatment can be given a course of 10 days followed by a rest of at least 15-20 days. There is no progression of dosage. Intensities of 1–2 W/cm^2 are used in areas that have a lot of muscle and less intensity is used for other areas (0.5–1.0 W/cm^2). For tendons or superficial tissue, 0.5 watts per square centimeter of head surface is used. For deeper penetration, 1 to 2 watts per square centimeter of head surface is used. The lower the frequency, the better will be the result. Duration of application varies between 5 to 10 minutes, and the head of the instrument should be kept in motion and in contact with the skin. Continuous ultrasound therapy is best used to heat tissues, and the pulsed ultrasound therapy is used when the non-thermal effects of ultrasound are desired. High doses of ultrasound will cause a rise in tissue heat to as high as 106 degree centigrade. The temperature rise can cause bone or tissue damage, and care should be taken not to use too high a dosage for too long a time. The head should be kept in motion to prevent heat from accumulating in the tissues underneath the head surface. An important principle of ultrasound therapy is that too small dosages/ intensities produce no useful effect, too much intensity causes tissue damage and the appropriate dosages provides desired result.

Ultrasound Dose Calculations

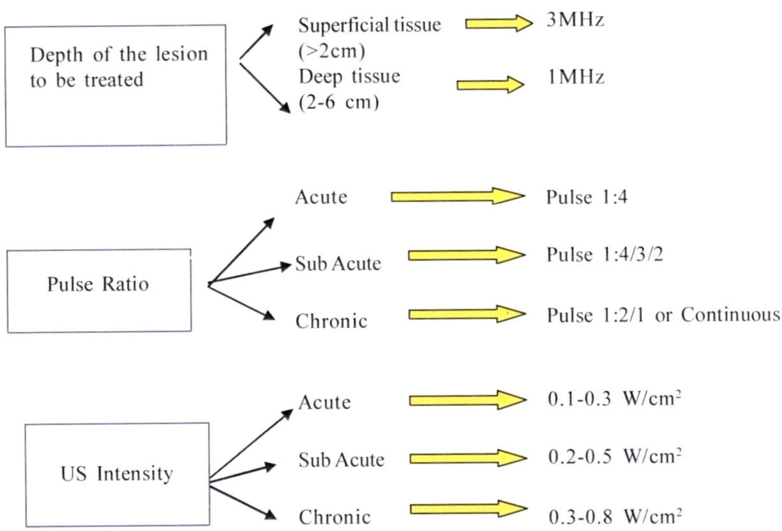

Treatment Time = 1x (No of times treatment head fits onto tissue to treat) x (pulse ratio)

Example : Ultrasound treatment for an acute lesion of the medial collateral ligament of the knee

- The lesion is superficial, hence a 3 MHz frequency would be most appropriate
- The lesion is acute, thus an intensity of 0.2-0.5 W/cm^2 should be sufficient
- The lesion is acute, therefore a pulse ratio of 1:4 is appropriate
- Using the large treatment head, it is estimated that the target tissue is approximately the same size as the treatment head (i.e. the head fits on to the tissue once)
- Working on the principle of 1 minutes worth of ultrasound per treatment head area, the total time taken to treat the lesion will be (1 minute) x (number of times the treatment head fits over the lesion) x (the pulse ratio)=(1) x (1) x (5)=5 minutes
- The final treatment dose will therefore be 3MHz, 0.2-0.5 W/cm^2, pulsed 1:4, treatment time -5 minutes.

Coupling Agent

Coupling media (most commonly gels) prevent reflection of the waves away of the soft tissue/air interface by excluding air between the transducer and skin surface. Different media have different impedance. Any coupling media should have acoustic impedance similar to that of the transducer, should absorb little of the ultrasound, remain free of air bubble and allow easy movement of the transducer over the skin surface.

A coupling agent is needed to connect the ultrasound head with the skin in order to maximize transfer of the ultrasound energy in top the tissues. Commercially different water-soluble gels are advisable in the market. Commercially available gel pads, water-filled latex gloves, gel-filled latex gloves, gel-filled condoms are also preferable if the direct method could not be used because of the contours of the treatment surface.

Phonophoresis

Phonophoresis refers to the use of ultrasound to enhance the delivery of topically applied drugs to the underlying tissues. A substance dissolved or suspended in a coupling medium can be penetrate through the skin into the underlying tissue up to 50 mm depth. The use of a continuous beam at a 2 Watt/cm^2 increases the permeability of the cell membrane by 200% and allows for absorption of the substance. The most common use of phonophoresis is for treatment of localized musculoskeletal inflammatory conditions. Hydrocortisone, dexamethasone, salicylates, indomethacin or lidocaine are incorporated into a vehicle such as glycerol, cream, oil or water for skin application. Ultrasound is then applied over the area in an effort to drive these substances into the tissues. The major limiting factor has been low skin permeability. Low-frequency phonophoresis can enhance the transdermal delivery of proteins such as insulin and interferon. Research in animals should be directed toward documenting drug penetration through the skin, and the effect of the hair coat.

Iontophoresis

Methods of transdermal drug delivery include iontophoresis and phonophoresis. All matter is composed of atoms that contain positively and negatively charged particles called ions. These charged particles possess electrical energy and can move about, propelled by an electrical force. Iontophoresis ("ion transfer") is a form of electrotherapy. Certain drugs, which ionize in solution, can be driven into the skin and underlying tissues by direct current applied through surface electrodes. Since an electrode will repel similarly charged ions, positively charged ions can be introduced into the tissues by the positive electrode (anode); the cathode can introduce ions with a negative charge.

Transcutaneous Electrical Nerve Stimulators (TENS)

Therapeutic electrical stimulators are often called TENS (Transcutaneous electrical nerve stimulators). It has pain-suppressing effect. TENS is a method of electrical stimulation, which primarily aims to provide a degree of pain relief (symptomatic) by specifically exciting sensory nerves and thereby stimulating either the pain gate mechanism and/or the Opioid system. The different methods of applying TENS relate to these different physiological mechanisms. Success is not guaranteed of applying TENS, and the percentage of patients who obtain pain relief is vary, but would typically be about 50-60% and it is better than the placebo effect. The technique is non invasive and has side effects when compared with drug therapy. Most TENS applications are now made using self-adhesive, pre gelled electrode. The small battery operated units supplied with two or four electrodes are designed to deliver a series of pulses current. The number of pulses per second (pps) and intensity can both be adjusted to inhibit pain perception. In this therapy, all therapeutic electricity is delivered through skin surface electrodes. TENS units block 'C' and 'A' delta fiber pain. The electrodes are placed on the skin (pain area) after applying liberal gel the electrodes, so that the adjusted current flow (as per manufacturer's instructions) passes through the pain source and produce an effective analgesic effect.

Indications

- Chronic pain
- Wound healing
- Soft tissue repair
- Osteoarthritis
- Edema reduction
- Vasodilation

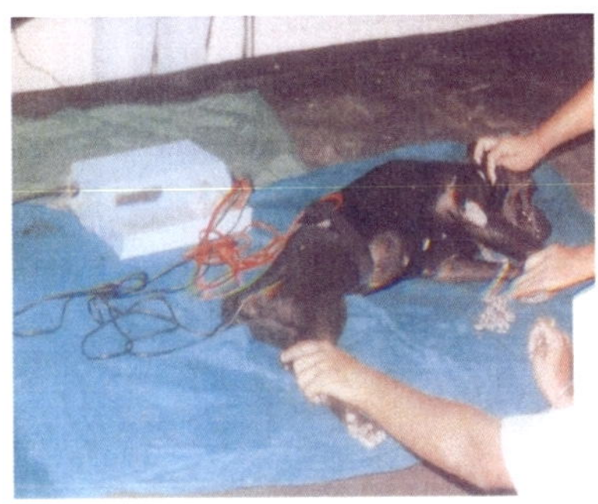

Contradications

- Open wound
- Infection/sepsis
- Non-cooperative patients
- Pregnancy
- Patient with a pacemaker
- Patients who have an allergic response to the electrodes, gel or tape
- Dermatitis
- Patients with recent bleeding/hemorrhage/thrombosis
- Application over the anterior aspect of the neck or carotid sinus
- Neoplasm (?)

Precautions

- Electrodes should not be placed over the eyes
- Patients who have epilepsy
- Avoid active epiphyseal regions in young animals
- If there is abnormal skin sensation, the electrodes should preferably be positioned in a comfortable site

Interferential Therapy (IFT)

Interferential therapy aids in the relief of pain and the promotion of soft tissue healing. Tiny amounts of electrical are induced into the tissues in the vicinity of the injury. Where these waves intersect below the surface of the skin, low frequency stimulation is created. This promotes the body to secrete endorphins and other natural painkillers to help relieve pain.

Indications

- Ligament sprains
- Muscle strains
- Muscle spasms
- Muscle injury
- Oedema
- Analgesia
- Muscular atrophy
- Haematoma
- Neuralgic pain
- Hindquarter weakness
- Paraplegia
- Spinal cord injury
- Urinary incontinence (nerve origin)
- Delayed union

Contradications

- Infection
- Open wounds
- Tumours
- Pregnancy
- The eyes
- The anterior aspect of the neck
- The carotid sinus
- Patients with pacemaker
- Dermatitis/broken skin
- Recent soft tissue injury
- Danger of hemorrhage/tissue bleeding
- Active epiphyseal region in young animals

- Patients taking anticoagulation therapy
- Patients have a history of pulmonary embolism
- Patients have a history of deep vein thrombosis

Precautions

- Patients are unable to co-operate should not be treated
- If there is abnormal skin sensation, electrodes should be positioned in a site other than this area to ensure effective stimulation

Hydrotherapy/Aquatic Therapy

In medical terms, hydrotherapy denotes treatment of disease by, or in, water. In its simplest form this could be bathing the pet animal using a commercially available herbal shampoo or using a stronger, medicated wash, to treat a specific skin condition. Aquatic therapy includes both swimming and other aquatic exercise. Aquatic exercise for small breed dogs is relatively easy compared to large breed dogs. Small dogs such as Dachshunds can be lifted into a sink or tub, and easily held for support. A heavier dog which is not ambulatory is much more difficult to transport, and will require a more sophisticated pool. The types of facilities that are used for aquatic therapy include

- Sinks
- Tubs
- Children wading pools
- Above ground pools
- Underwater treadmill systems
- Ponds, lakes and beaches

Indications

- Orthopedic problems: hip dysplasia, cruciate rupture, post-surgical
- Neurological problems: paralysis, weakness, disc herniations, degenerative myelopathy, etc.
- Obesity
- De-conditioned ("out-of-shape") pets.

Swimming

This is a form of rehabilitation therapy particularly appropriate for dogs and horses recovering from a range of conditions - especially those involving problems with limbs. This is a non-weight bearing exercise in which the limbs work against the resistance of the water in a symmetrical pattern. It should be used whenever possible to strengthen the limbs after surgery before regular training procedures are reinstituted. The animal should swim in straight lines. When animal shows the signs of exhaustion, swimming should be stopped. Swimming for the first circuit should be short followed by a period of rest and then start for second circuit of swimming. Heart and respiration should return normal before the start of second circuit. All animals can swim by instinct. How much they enjoy it depends on their confidence in the water and this can be increased by gradual acclimatization.

Benefits of hydrotherapy/Swimming

Swimming helps to improve general fitness and stamina. Improves muscle tone and is helpful in recovering from injury or operation. Muscle wastage begins within 3 days of any immobilization so to prevent further weakness or injury. It is important to rebuild, through safe exercise, any muscles that have deteriorated. It is better to swim dogs in heated water since cold water causes constriction of the blood vessels near the skin and to the superficial muscles, which restricts the flow of blood making the muscles less efficient. However, horses are usually swum in cold water because it helps to dissipate the enormous amounts of heat they generate under exertion. This is far more than dogs and with a less efficient cooling system. Therefore heated pools and cold-water pools are more useful for dogs and horses respectively. Swimming is an excellent form of exercise; most of muscles normally used in movement are involved. Because of the increased resistance to movement, the muscles have to work harder than they would do on land. It improves the general fitness of the cardiovascular and respiratory system of the animal. Breathing requires more effort in water-chest and abdominal muscle strength.

Contraindications

- Old animal
- Unfit animal

[Some text of this chapter is extracted from our Book "Restraint, First Aid and Physiotherapy in Animals" by Chakrabarti A, Nandi SK and Maiti SK.].

References / Suggested Reading

Albee F.H. (1944). Evolution of bone graft surgery. American Journal of Surgery, 63: 421-436.

Aldinger G., Kusswetter G.H.W., Reis H.J. *et al.* (1991). Bone morphogenetic protein: a review, Internatonal Orthopaedics 15: 169-177.

Albrktsson T. (1980). Repair of bone grafts. A vital microscopic and histological investigation in the rabbit. Scand J Plast Reconstr Surg, 14: 1-12.

Asahina I., Sampath T.K. and Hauschka P.V. (1996). Exp Cell Res 222: 38.

Bajpai P.K. (1983). Biomaterials in Reconstructive Surgery. Rubin LR (Ed), C V Mosby, St Louis, pp. 312-328.

Barger Lux M.J. and Heaney R.P. (1993). Effects of calcium restriction on metabolic characteristics of premenopausal women. J Clin Endocrinology Metabolism. 76: 103-107.

Bascom C.C., Sipes N.J., Coffey R.J. and Moses H.L. (1989). J. Cell Biochem 39: 25-32.

Bhadane Bhagyashree K., Maiti S.K., Divya Mohan, Bag S. and Kumar N. (2018). Role of embryonic stem cell-hydroxyapatite construct with growth proteins for osteogenesis in the repair if bone defects in rabbit model. Journal of Stem Cell Research & Therapeutics, 4(4): 67-80 DOI: 10.15406/jsrt.2018.04.001117

Bhasker, S.N., Brady J.M., Getter L., Greower M.F. and Driskell T.D. (1971). Biodegradable ceramic implants in bone. Oral Surgery 32: 336-346.

Brinker W.O. (1948). The use of intramedullary pins in small animal fracture. North Am Vet 29: 292.

Brinker W.O. (1978). Small Animal Fracture. East Lansing, Michigan. Department of Continuing Education Service. Small Animal fracture, Michigan state University Press.

Bromiley M.W. (1991). Physiotherapy in Veterinary Medicine. Oxford: Blackwell Scientific Publications.

Boyne P.J. (1973). Induction of bone repair by various bone grafting materials. In: Hard Tissue Growth, Repair, Reconstruction. Associated Scientific Publishers, Amsterdam, pp. 121-141.

Brown K.L. and Cruess R.L. (1982). Bone and cartilage transplantation in orthopaedic surgery: a review. L Bone & Joint Surg, 64: 270-279.

Burwell R.G. (1961). A study of homologous cancellous bone combined with autologous red marrow after transplantation to a muscular site. J Anat 95: 613-616.

Burwell R.G. (1969). The fate of bone grafts. In: Recent Advances in Orthopaedics, Apley AG (Ed), Baltimore, Williams and Wilkins, pp. 115-207.

Buckwalter J.A., Glimcher M.J., Cooper R.R. and Recker R. (1995). Bone Biology. J Bone & Joint Surg Am. 77-A (8): 1256- 1275.

Burchardt H. (1983). The biology of bone graft repair. Clin Orthop. 174: 28-42.

Burcharrdt H. (1987). Biology of bone transplantation. Orthop Clin North Am. 18: 187-196

Calvo M.S., Kumar R. and Heath H. (1990). Persistently elevated parathyroid hormone secretion and action in young women after four weeks of ingesting high phosphorus, low calcium diets. J Clin Endocrinology Metabolism, 70: 1334-1340.

Cameron Hu, Macnab I. and Piliar R.M. (1977). Evaluation of biodegradable ceramics. J Biomed Mater Res 11: 179.

Campbell J.R. and Douglas T.A. (1965). The effect of low calcium intake and vitamin D supplements on bone structures in young growing dogs. Br J Nutrition. 19: 339.

Canalis E. (1983). Endocrinology Review 4: 62-63.

Centrella M., McCarthy T.L. and Canalis E.J. (1991). J Bone and Joint Surgery A73: 1418-1428.

Centrella M., McCarthgy T.L. and Canalis E. (1992). Bone Metabolism and Mineralisatation BK Hall (Ed) CRC Press, Boca Raton 4: 47-72.

Celeste A.J., Iannazzi J.A., Taylor R.C. *et al.* (1990). Proc Natl Acad Sci USA. 87: 9843-9847.

Chakrabarti A., Nandi S.K. and Maiti S.K. (2007). Restraint, First aid and Physiotherapy in animals. Kalyani Publishers, Ludhiana, New Delhi.

Chapman W.L. (1965). Appearance of ossification centers and epiphyseal closure as determined by radiographic techniques. J Am Vet Med Assoc. 147: 138.

Chaturvedi S.N., Singh K.N., Shukla R.P. and Singh P.K. (1980). Effect of zinc supplementation on fracture healing. Ind J Orthopaedics. 14: 62-65.

Choudhari P. and Baxi M. (2016). Minimally Invasive Plate Osteosynthesis. Indian J Ortho Surg. 2: 194-198.

Conzemius M. and Swainson S. (1999). Fracture fixation with screws and bone plates. Veterinary Clinics North America: Small Animal Practice 29 (5): 1117-1133.

Crenshaw A.H. (1987). Campbell's Operative Orthopaedics, edn 7, CV Mosby, St Louis.

Cummings R.G. (1990). Calcium intake and bone mass: A quantitative review of the evidence. Calcif Tissue Inter. 47: 194-201.

Denny H. (1986). Acute trauma in small animals. 3. Orthopaedic injuries. In Practice 8: 168-175.

Denny H.R. (1991). Fracture fixation in small animal practice. In Practice13: 137-143.

Daculsi G., Passuti N. and Martin S. (1990). Macroporous calcium phosphate ceramic for long bone surgery in humans and dogs. J Biomed Meter Research 24: 379-396.

De Groot K. (1980). Bio-ceramics consisting of calcium phosphate salts. Biomaterials 1: 47-50.

Deyoung D.J. and Probst C.W. (1985). Methods of fracture fixation. Textbook of Small Animal Surgery, Slatter DH (ed) WB Saunders, Philadelphia, pp. 1949-1972.

Downer A.H. (1978). Physical Therapy for Animals. Selected Techniques. Springfield: Charles C Thomas. 63-74.

Ehner E.A. (1947). J Bone pinning in fractures of small animal Am Vet Med Assoc 110: 14.

Einhorn T.A. (1995). J Bone and Joint Surgery 77A: 940.

Fazili M.R., Chawala S.K., Tayal R., Behl S.M. and Singh K. (2005). Physiotherapy in Veterinary Practice. Intas Polivet 6: 1-10.

Ferguson D., Davis W.L., Urist M.R., Hurt W.C., Allen E.P. (1987). Bovine bone morphogenetic protein (bBMP) Fraction induced repair of craniotomy defects in the rhesus monkey (Macaca speciosa). Clin Orthop 219: 251-258.

Frirdlaender G.E. (1987). Current concept review: Bone grafts. The basic science rationales for clinical applications. L Bone Joint Surg. 69: 786-790.

Gahlot T.K. and Chouhan D.S. (1988). Roentgenological evaluation of fracture healing following uses of anabolic hormone and vitamin C in dogs. Ind J Anim Sci, 58: 1196-1199.

Graves G.A., Hentrich R.L., Stein H.G., Bajpai P.K. (1971). Resorbable ceramic implants. J Biomed Meter Res Symp 2 (part I) 91-115.

Gross D.M. (2002). Canine Physical Therapy: Orthopedic Physical Therapy. Wizartd of Paws: East Lyme, Connecticut. 315.

Habal M.B. and Reddi A.H. (1992). Bone grafts and bone substitutes. WB Saunders Company, Philadelphia, pp. 6-52.

Hamadouche M. and Sedel L. (2000). Ceramics in Orthopaedics. J Bone Joint Surgery 82-B 1095-1099.

Hartley W.J., Karter J.C. and Mackay A. (1963). Goiter and low copper status in a litter of meat fed pups. New Zealand Vet J. 11: 1.

Heaney R.P. (1993). Nutritional factors in osteoporosis. Annual Review of Nutrition. 13: 287-316.

Heppenstall R.B. (1980). Bone grafting. In: Fracture treatment and healing. Happenstall RB (Ed) WB Saunders Company, Philadelphia, pp. 97-112.

Heaney R.P. and Recker R.R. (1982). Effects of nitrogen, phosphorus and caffeine balance in woman. J Lab Clin Med. 99: 46-55.

Hinko P.J. and Rhinelander F.W. (1975). Effective use of cerclage in the treatment of long bone fracture in the dog. J Vet Med Assoc 166: 520.

Hudson C.C., Antonio P., Lewis D.D. (2009). Minimally invasive plate osteosynthesis: Applications and techniques in dogs and cats. Veterinary and Comparative Orthopaedics and Traumatology 22(3): 175-82.

Isobe M. (1995). Kokubyo Gakkai Zasshi 62: 506.

Jones Gary Clayton (1994). Failure of fracture repair. In Practice, September, http://dx.doi.org/10.1136/inpract.16.5.256

Johnson K.A. (1980). The effects of autologous bone graft on bone healing after carpal arthrodesis in the dog. Vet Rec 107: 126-132.

Joyce M.E., Roberts A.B., Sporn M.B. *et al.* (1990) J Cell Biology. 110: 2195-2207.

Kanatani M., Sugimoto T., Kaji H. *et al.* (1995). J Bone miner Res. 10: 1681.

Koskinen *et al.* (1971). Bone formation by induction under the influence of growth hormone and cortisone. Israel J Med Res, 7: 378-380.

Krieger N.S., Sessler N.E. and Bushhinsky D.A. (1992). Acidosis inhibits osteoblastic and stimulates osteoclastic activity in vitro. Am J Physol. 262: F 442-F 448.

Le Geros R. and Le Geros J. (1993). Dense hydroxyapatite In: An introduction to Bioceramics, Hence Li and Wilson J (Eds), World Scientific Publishing Co, Singapore, pp. 139-180.

Leman J. Jr, Pleuss J.A. and Grany R.W. (1993). Potassium causes calcium retention in healthy adults. Journal of Nutrition. 123: 1623-1626.

Maiti S.K., Ninu A.R., Sangeetha P., Mathew D.D., Tamilmahan P., Kritaniya D., Kumar N. and Hescheler J. (2016). Mesenchymal stem cells-seeded bioceramic construct for bone regeneration in large critical-size bone defect in rabbit. Journal of Stem Cells & Regenerative Medicine (Pubmed), 12(2): 87-99.

Maiti S.K., Shivakumar M.U., Mohan Divya, Kumar N., and Singh K.P. (2018). Mesenchymal stem cells of different origin-seeded bioceramic construct in regeneration of bone defect in rabbit. Tissue Engineering and Regenerative Medicine (Springer) 15(4): 477-492; DOI 10.1007/s13770-018-0129-7

Maiti S.K., Naveen Kumar, Singh, G.R. and Hoque M. (2003). Effect of Ultrasound therapy on fracture healing in rabbits. Journal of Applied Animal Research 24: 25-32.

Maiti S.K. (2001). Nutritional management of fracture healing. Short course on "Nutrition and Dietetics under clinico-therapeutic conditions of pet and farm animals" (March 20- April 18) Division of Animal Nutrition, IVRI, Izatnagar (UP).

Maiti S.K. (2003). Internal fracture fixation techniques in animals. Advance Training Course on "Current Advances in Veterinary Surgery". Division of Surgery, ICAR-Indian Veterinary Research Institute, Izatnagar (UP).

Maiti S.K. (2003). Role of different minerals, vitamins and hormones in fracture healing of dairy animals. ICAR Summer School on "Advanced diagnostic techniques and therapeutic approaches to metabolic and deficiency diseases in dairy animals". (July 15-August 4), Division of Medicine, IVRI, Izatnagar (UP).

Maiti S.K. and Divya Mohan (2015). Bone grafts and bone substitutes in orthopaedic surgery. Short Course on "Fracture Management in Animals" (November-December), Surgery Division, ICAR-Indian Veterinary Research Institute, Izatnagar (UP).

Maiti S.K. and Divya Mohan (2019). Fracture types and fracture healing. Short Course on "Fracture Management in Animals" (1-10-2019-31-10-2019). Division of Surgery, ICAR-IVRI, Izatnagar (UP).

Maiti S.K. (2019). Minimally invasive fracture repair techniques and biological osteosynthesis. Short Course on "Fracture Management in Animals" (1-10-2019-31-10-2019). Division of Surgery, ICAR-IVRI, Izatnagar (UP).

Maiti S.K. and Singh G.R. (1996). Recombined bone grafts and ceramic implants with bone marrow and plasma in goats: Gross evaluation. Journal of Applied Animal Research, 9: 193-200.

Maiti S.K. and Singh G.R. (1996). Studies on vascularisation and ceramic implants in goats. Small Ruminant Research, 20: 171-176.

Maiti S.K., Kalicharan and Singh G.R. (1995). A new staining approach for evaluation of different types of bone grafts and ceramic implants in goats. Indian Journal of Animal Sciences, 65: 751-753.

Maiti S.K. and Singh G.R. (1995). Radiological evaluation of composite bone grafts and ceramic implants in goats. Indian Journal of Animal Science, 65: 1182-1188.

Maiti S.K., Kalicharan and Singh G.R. (1995). Histopathological evaluation of composite bone grafts and ceramic implants in goats. Indian Veterinary Journal, 72: 728-733.

Maiti S.K. and Singh G.R. (1995). Undecalcified ground section study in composite bone grafts and ceramic implants in goats. Indian Journal of Veterinary Surgery 16: 30-34.

Maiti S.K. and Singh G.R. (1995). Different types of bone grafts and ceramic implants in goats: A triple fluorochrome labeling study. Indian Journal of Animal Sciences, 65: 140-143.

Maiti S.K., Aithal H.P., Hoque M. and Singh G.R. (1999). Acceleration of bone healing with progesterone supplement-a radiological and scanning electron microscopic study. Intas Polivet II (2): 106-109.

Maiti S.K., Hoque M., Aithal H.P. and Singh G.R. (1999). Repair of experimental fractures under influence of anabolic steroid: clinico-radiological study. Intas Polivet, II (2): 129-131.

Maiti S.K., Shiva Kumar M.U., Srivastava L., Kumar N., Ninu A.R. (2012). Isolation, proliferation and morphological characteristics of bone marrow derived mesenchymal stem cells (BM-MSC) from different animal species. Trends Biomater Artif Org 2012; 27(1): 29-35.

Maiti S.K., Singh Kiranjeet, Gopinathan A. (2019). Application of plaster cast/splint for fracture immobilization and other bandaging procedures in large and small animals, Internship (UG) Manual, ICAR-IVRI Deemed University, Izatnagar (UP).

Maiti S.K. (2012). Isolation, proliferation and transplantation of bone marrow derived animal mesenchymal stem cells in critical sized bone defects in animal model. Journal of Biotechnology and Biomaterials, 2(6): 143.

Maiti S.K. and Singh G.R. (1998). Bone morphogenetic proteins (BMP) - Novel regulators of bone formation: A perspective. Indian Journal of Experimental Biology, 36: 237-244.

Maiti S.K. and Singh G.R. (2003). Ceramic Biomaterials in reconstructive surgery-A review. Indian Journal of Veterinary Surgery 24: 1-10.

Maiti S.K. (2002). Internal fracture fixation in small animal orthopedics-A perspective. Intas Polivet. 3: 307-315.

Maiti S.K., Hoque M., Aithal H.P., Singh G.R. and Kalicharan (2002). Effect of non-steroid anti-inflammatory drug (NSAID) on fracture healing: An experimental study in rabbits Indian Veterinary Medical Journal 26: 331-334.

Maiti S.K., Saravanan B., Singh G.R., Kumar N., Hoque M., Jawaharlal and Kalicharan (2007). Evaluation of the herb, *Cissus quadrangularis* in accelerating the healing process of femur osteotomies in dogs. Journal of Applied Animal Research 31: 47-52.

Maiti S.K., Singh G.R. and Mogha I.V. (2002). Bone Allograft – A review. Indian Journal of Veterinary Surgery 23: 1-11.

Maiti S.K., Dutta A., Singh G.R., Kumar N and Hoque M. (2003). Effect of herbal Magacal on fracture healing. Phytomedica 4: 1-12.

Maiti S.K., Hoque M., Aithal H.P. and Singh G.R. (1999). Repair of experimental fracture under influence of anabolic steroid: Clinico-radiological study. Intas Polivet, II: 129 -131.

Maiti S.K., Hoque M., Aithal H.P., Singh G.R. and Kalicharan (2003). Effect of bone wax on fracture healing: An experimental study in rabbits. Indian Journal of Animal Sciences, 73: 278-280.

Maiti S.K., Hoque M., Aithal H.P., Singh G.R. and Kalicharan (2002). Effects of gonadotropic hormone (Gn-RH) on fracture healing. Indian Journal of Animal Sciences, 72: 47-49.

Maiti S.K. and Singh G.R. (1996). Transforming growth factor-beta (TGF-â) in bone remodeling. Current Science, 71: 613-617.

Maiti S.K., Hoque M., Aithal H.P., Singh G.R. and Kalicharan (2001). Effects of prostaglandin on fracture healing an expt. Study on rabbits. Indian Journal of Animal Sciences, 71(8): 745-748.

Maiti S.K. (2008). Current status on diagnosis and management of fractures in animals. Short Term Training Course on Laboratory Diagnosis of Animal Diseases and Zoonoses. ICAR-IVRI Regional station, Kolkata. pp: 141-148.

Maiti S.K., Sharma A., Kumar N., Gupta O.P. and Sharma A.K. (2007). Effects of ultrasound and interferential therapies on hindquarter weakness in dogs. Indian Journal of Animal Science, 77: 1273-1276.

Maiti S.K., Naveen Kumar, Singh G.R., Hoque M. and Rajendra Singh (2006). Effect of ultrasound therapy on tendon healing in goat. Journal of Veterinary Medicine, Series-A 53: 249-258.

Maiti S.K., Pradhan P., Kumar N., Singh G.R. and Singh R. (2006). Effect of therapeutic ultrasound in acute traumatic arthritis in rabbits. Indian J. Animal Sciences, 76: 292-295.

Maiti S.K., Naveen Kumar, Singh G.R., Pawde A.M., Hoque M. and Rajendra Singh (2004). Effect of ultrasound therapy on muscular injury in rabbit. J. Applied Animal Research 26: 67-72.

Maiti S.K. (2020). "Stem Cell Research in Lab Animals". New India Publishing Agency, New Delhi.

Mattie D.R., McFall F.B., Bajpai P.K. and Gridely J. (1984). ALCAP ceramics: a non-toxic aluminium, calcium and phosphorus oxide biomaterials. Fed Proc 43: 327.

Martinoz S.A. and Walker T. (1999). Bone graft. Vet Clin North Am Small Anim Pract. 29: 1207-1219.

Marsell R. and Einhorn T.A. (2011). The biology of fracture repair. Injury, 42(6): 551-555.

Martens Ann, Pader K., Prange T., Ortved K.F. (2019). Minimally Invasive Surgical Techniques, (Chapter-13), In Equine Surgery (5th Ed) (Ed: Jorg A Auer and John A Stick) Elsevier.

McDowedl L.R. (1992). Minerals in animal and human nutrition. Academic Press. New York.

McLaughlin R.M. and Roush J.K. (1999). Repairing fractures with bone plate and screw fixation. Vet Med 94: 64-72.

Muller ME *et al.* (1979). Manual of Internal Fixation, Springer, Verlag, Berlin

Mustoe T.A., Pierce Gf, Thomason A. *et al.* (1987). Science. 237: 1333-1336.

Nielsen H.M., Andreassen T.T., Ledet T. *et al.* (1994). Acta Ortop Scandinavica. 65: 37-41.

Ninu A.R., Maiti S.K., Sangeetha P., Kritaniya D., Tamil M. and Kumar N. (2016). *In vitro* osteoinduction potential of a novel silica coated hydroxyapatite bioscaffold seeded with rabbit mesenchymal stem cell. Journal of Stem Cell Research and Therapeutics, 1(2): 49-55.

Ninu A.R., Maiti S.K., Shiva Kumar M.U., Sandeep Kumar, Sangeetha P., Deepika K., Gupta S., Saxena A. and Kumar N (2017) Isolation, proliferation, characterization and in vivo osteogenic potential of bone marrow derived mesenchymal stem cells (rBMSC) in rabbit model. Indian Journal of Experimental Biology, 55 (2):79-87.

Noda M. and Camilliere J.J. (1989). Endocrinology. 124: 2991-2994.

Nunamaker D.M. (1985). Methods of internal fixation. Textbook of Small Animal Orthopedics, JB Lippincott Co, Philadelphia, pp. 261-286.

Pandey S.K. (1985). Influence of certain hormones on histological and biochemical changes in callus formation in dogs. Ind J Anim Sci. 55: 527-531.

Price P.A. (1985). Vitamin K-dependant formation of bone Gla protein and its function. Vitamins and Hormones. 42: 65-108.

Porter M. (1998). Physical therapy, In : Complementary and Alternative Veterinary Medicine, Schoen AM and Wynn SG (Eds), Mosby, St Luis.

Radheysham (2012). Slideshow: Seminar on Principles and Techniques of AO, Department of Orthopaedics, Gandhi Medical College, Bhopal, (MP), India, Internet.

Raisz L.G. *et al.* (1978) Hormonal regulation of bone formation. Recent Progress in hormone Research (Vol 34). Academic Press, pp. 335-356.

Reddi A.H. and Huggins C. (1972). Proc Natl Acad Sci, USA, 69: 1601.

Richardson Dean W. (2008). Less invasive techniques for equine fracture repair and arthrodesis. Vet Clin Equine 24: 177-189.

Richardson D.W., Ortved K.F. (2019). Minimally Invasive Fracture Repair, In Equine Surgery [5th Ed] e-book (Ed: JA Auer and JA Stick) Chapter: 13: p. 229-232.

Rosen H. (1975). Principles and application of bone plates. Vet Clin North Am 5: 299.

Rudy R.L. (1975). Principles of intramedullary pinning. Vet Clin North Am 5: 209.

Roush J.K. and McLaughlin R.M. (1978). Principles of applying pin and wire fixation. Vet Med 93: 1075-1080.

Roush J.K. and McLaughlin R.M. (1988). Fundamentals of fracture management. Vet Med 93: 1065-1070.

Saravanan B., Maiti S.K., G.R. Singh, M. Hoque, N. Kumar and H.C. Setia (2004). Complications associated with different internal fixation techniques in the management of comminuted femoral fracture in dogs. Indian Journal of Animal Sciences 74: 128-129.

Saravanan B., Maiti S.K., Hoque M., Aithal H.P. and Singh G.R. (2002). Management of comminuted femoral fracture by different internal fixation techniques in dogs. Indian Journal of Animal Sciences, 72: 1104-1107.

Saravanan B., Maiti S.K., Hoque M., Aithal H.P. and Singh G.R. (2002). Histopathological evaluation of different internal fixation techniques in the management of comminuted femur fracture by in dogs. Indian Journal of Animal Sciences 72: 665-667.

Saravanan B., Maiti S.K., Hoque M., Naveen Kumar and Singh G.R. (2003). Evaluation of different internal fixation techniques for the management of comminuted femoral fracture in dog: A scanning electron microscopic (SEM) study. Indian Journal of Animal Sciences, 73: 399-401.

Schwartz R. and Reddi A. (1979). Influence of magnesium depletion on matrix-induced endochondral bone formation. Calcif Tissue Inter. 29: 15-20.

Sharma A.K., Kumar Naveen and Maiti S.K. (2007). Orthopaedic Biomaterials, Division of Surgery, Indian Veterinary Research Institute, Izatnagar (UP).

Singh R.H. and Udupa K.N. (1963). Certain biological studies on the effect of cortisone in healing of fractures. Ind J Med Res. 51: 1028.

Singh R.H. and Udupa K.N. (1966). Some investigations on the effect of insulin in healing of fractures. 54: 1071-1082.

Singh R.H. and Udupa K.N. (1966). Studies on the effect of parathormone and vitamin D in healing of fractures. Ind J Med Res. 54: 872-880.

Snow G.R. and Anderson C. (1985). The effects of continuous progesterone treatment on cortical bone remodelling activity in Beagles. Calcif Tissue Int. 37: 282-286.

Snow G.R. and Anderson C. (1986). The effects of 17 beta-Estradiol and progesterone on trabecular bone remodelling in dogs. Calcif Tissue Int. 39: 195-205.

Stevenson S. (1993). Bone grafting. In: Textbook of Small Animal Surgery. Slatter DH (Ed), WB Saunders Company, Philadelphia, pp 1694-1703.

Steiss J.E. (2000). Physical therapy in veterinary medicine. Therapeutic ultrasound and phonophoresis. Comp Cont Edu Pract Vet, 22: 690-693.

Stuart Julie (2005). New England Physical Therapy for Animal (NEPTA) (www.pt4animals.com)

Sumner D.R., Turner T.M., Purchio A.F. *et al.* (1995). J Bone Joint Surgery A. 77: 1135-1148.

Sulong Ahmad Fadzli (2017). Slide Share: Locking Compression Plate (LCP) Principle, International Islamic University, Malaysia.

Suryawanshi S.B., Maiti S.K. and Kesava Rao (1999). Effect of endocrine factors on callus composition in experimental tibial metaphyseal fracture in dogs. Indian Veterinary Journal, 76: 125-127.

Suryawanshi S.B., Maiti S.K. and Singh G.R. (2000). The effect of thyroxin and insulin hormones in fracture healing-radiological study. Indian Journal of Animal Sciences, 70: 171-173.

Suryawanshi S.B., Maiti S.K. and Singh G.R. (2001). Repair of tibial metaphyseal fractures by intramedullary pinning in dogs-A ground study. Indian Journal of Animal Sciences, 71(4): 355-358.

Suryawanshi S.B., Maiti S.K. and Singh G.R. (2001). A comparative study of two different fixation techniques in the management of tibial metaphyseal fracture in dogs. Journal of Applied Animal Research, 20: 89-98.

Suryawanshi S.B., Maiti S.K. and Somvanshi R. (1999). Histopathological evaluation of thyroxin and insulin therapy for the management of tibial metaphyseal fracture in dogs. Indian J of Animal Sciences, 69: 116-117.

Suryawanshi S.B., Maiti S.K., Singh G.R. and Varshney V.P. (1999). Effect of anabolic hormones on haemato-biochemical changes in fracture healing in dogs. Indian Journal of Animal Sciences, 69: 404-406.

Suryawanshi S.B., Maiti S.K. and Singh G.R. (1999). Fluorochrome labeling studies on the effect of anabolic hormones in fracture healing in dogs. Indian Journal of Veterinary Surgery, 20: 28 -30.

Takagi K. and Urist M.R. (1982). Clin Orthop, 171: 224.

Taylor Robert (2001). Physical therapy in veterinary medicine. WSAVA Congress, Vancouver.

Udupa K.N. and Gupta L.P. (1965). The effect of growth hormone and thyroxine in healing of fracture. 53: 7.

Underwood E.J. (1977). Trace elements in human and animal nutrition. Academic Press, New York.

Underwood E.J. (1982). Mineral nutrition in Livestock, CAB, London.

Uchida A., Nade S.M., McCartney E.R. and Ching W. (1984). The use of ceramics for bone replacement. A comparative study of three different porous ceramics. J Bone Joint Surgery 66B: 269-275.

Urist M.R. (1965). Bone formation by autoinduction. Science 150: 893-899.

Urist M.R., Silverman Bf, Buring K., Dubac Fl, Rosenberg J.M. (1967). The bone induction principle. Clin Orthop 53: 243-283.

Urist M.R. and Strates B.S. (1971). Bone morphogenic protein. J Dent Res, 50: 1392-1406.

Urist M.R. and Iwata H. (1973). Preservation and biodegradation of the morphogenetic property of bone matrix. J Theo Biol 38: 155.

Urist M.R., Mikulski A.J. and Lietze A. (1979). Proc Natl Acad Sci, USA 76: 1828.

Urist M.R., Mikulski A.J. (1979). Proc Soc Exp Biol Med 162: 48.

Urist M.R., Sato K., Brownell Ag *et al.* (1983). Human bone morphogenetic protein. Proc Sco Expt Biol Med, 173: 194-199.

Venugopalan A. (2002). Essential of Veterinary Surgery (8th Edn) Oxford & IBH Publishing Co Pvt. Ltd., New Delhi.

Wagner Michael, A.T. (2016). Slide Show: Fracture Fixation using the locking compression plate, Internet.

Wang Ea, Rosen V., Cordes P., Hewick R.M. *et al.* (1988). Proc Natl Acad Sci, 85: 9484.

Weiland A.J., Moore J.R. and Daniel R.K. (1983). Vascularized bone autograft: experience with 41 cases. Clin Orthop, 174: 87-95.

Wolf EF (1975). Rush pins in veterinary Orthopaedics-a review. Am Anim Hosp Assoc 11: 756.

Yang C., Simmons D.J. and Lozano R. (1994). The healing of grafts combining freeze-dried and demineralized allogenic bone in rabbits. Clinical Orthopedics Related Research 298(1): 286-95.